D0558432

Encountering Earth

Encountering Earth

Thinking Theologically With a More-Than-Human World

Trevor George Hunsberger Bechtel,
Matthew Eaton,
& Timothy Harvie

CASCADE *Books* · Eugene, Oregon

ENCOUNTERING EARTH
Thinking Theologically With a More-Than-Human World

Cascade Books
An Imprint of Wipf and Stock Publishers
199 W. 8th Ave., Suite 3
Eugene, OR 97401

www.wipfandstock.com

PAPERBACK ISBN: 978-1-4982-9784-4
HARDCOVER ISBN: 978-1-4982-9786-8
EBOOK ISBN: 978-1-4982-9785-1

Cataloguing-in-Publication data:

Names: Bechtel, Trevor George Hunsberger, Matthew Eaton, and Timothy Harvie, eds.

Title: Encountering Earth : thinking theologically with a more-than-human world / Trevor George Hunsberger Bechtel, Matthew Eaton, and Timothy Harvie.

Description: Eugene, OR: Cascade Books, 2018 | Includes bibliographical references and index.

Identifiers: ISBN 978-1-4982-9784-4 (paperback) | ISBN 978-1-4982-9786-8 (hardcover) | ISBN 978-1-4982-9785-1 (ebook)

Subjects: LCSH: Philosophy of nature. | Human ecology. | Theology—Christian. | Title.

Classification: BD581 .E47 2018 (paperback) | BD581 (ebook)

Manufactured in the U.S.A. 05/09/18

To Bones, Fargo, and Tiamat,
without whom this volume would not exist.

Contents

Contributors

Trevor George Hunsberger Bechtel

Trevor Bechtel is the Creative Director of Anabaptist Bestiary Project, a rock and roll band that sings about how God's will for human life is revealed in creation. They released their second CD *Of Every Creature* on 4action Records in 2013 (see anabaptistbestiaryproject.com). His research interests include the human/animal relationship, biblical interpretation, and the theology of computing and social media. Over the last two decades he has served on the faculty and administration of several small colleges in the Great Lakes region. He preaches and presents on these topics frequently throughout the Great Lakes. He published *The Gift of Ethics* (Eugene, OR: Cascade, 2014). An ordained minister in Mennonite Church, Bechtel lives in Ann Arbor, Michigan with Susan Hunsberger and Neko the Cat.

John Berkman

John Berkman is Professor of Moral Theology at Regis College, University of Toronto. In 2017 he was Visiting Research Scholar at the McDonald Centre at Christ Church College, Oxford and at the Aquinas Institute at Blackfriars, Oxford. He has been reflecting theologically on nonhuman animals for twenty-five years. He began with "A Trinitarian Theology of the Chief End of All Flesh" (with Stanley Hauerwas) in 1992, and recently published "Just Chimpanzees? A Thomistic Perspective on Ethics in a Nonhuman Species" in *Beastly Morality*, edited by Jonathan Crane (New York: Columbia University Press, 2015). In 2014 he organized and coedited the first volume of essays in Catholic moral theology dedicated to the moral analysis of the relationship between human and nonhuman animals. He lectures internationally to many groups on theology and nonhuman animals. Some of his many essays on nonhuman animals, as well as other topics, can be found on his web page on academia.edu. For many years Berkman shared his home with Thelma and Louise, two calico-colored feline siblings. Berkman lives in Toronto, Canada with his wife Jennifer and their three children. In 2018, God willing, a Newfoundland dog will join the family.

Kimberly Carfore

Kimberly Carfore is a doctoral candidate in Ecology, Spirituality, and Religion, and teaches Nature Immersion at the University of San Francisco. She was a research assistant for the Forum on Religion and Ecology at Yale University, is currently the Ecology and Religion Chair and Student Representative for the American Academy of Religion, Western Region, and has published multiple essays on the topics of ecofeminism, deconstruction, ecopsychology, and globalization. With a former career as a wilderness therapy instructor for the School of Urban Wilderness Survival of Idaho and North Carolina, and background in psychology, she enjoys thinking and being outside, painting and photography, and is an avid runner.

Lisa E. Dahill

Lisa E. Dahill is Associate Professor of Religion at California Lutheran University, Thousand Oaks, California. A native Californian, she is happy to be back within biking distance of the Pacific Ocean, although she misses the creeks and rivers of central Ohio, where she taught for ten years (Trinity Lutheran Seminary, Columbus, Ohio). Her publications explore the spirituality and writings of Dietrich Bonhoeffer, feminist models of spiritual formation, liturgical theology, and ecological implications of the Christian sacraments. She has chaired two American Academy of Religion groups and is past president of the Society for the Study of Christian Spirituality.

Celia Deane-Drummond

Celia Deane-Drummond is Professor in Theology and Director of the Center for Theology, Science and Human Flourishing at the University of Notre Dame. She holds a masters degree in natural science (Cambridge), a doctorate in plant physiology (Reading), an honors degree in theology (Bristol, CNAA) and a second doctorate in systematic theology (Manchester) and a postgraduate certificate in education (Manchester). Her interests are in the engagement of theology and theological ethics with natural/biological sciences, including evolution, ecology, psychology, animal behavior, anthropology, and environmental ethics. She is Chair of the European Forum for the Study of Religion and Environment (2011–present). She has published 200 peer-reviewed articles or book chapters and twenty-five books either as author or editor. She is joint editor of the international journal *Philosophy, Theology and the Sciences*. Recent books include *Wonder and Wisdom* (London: DLT, 2006), *Genetics and Christian Ethics* (Cambridge: Cambridge University Press, 2006), *Future Perfect*, edited with Peter Scott (New York: Continuum, 2006, 2010), *Ecotheology* (London: DLT, 2008), *Christ and Evolution* (Minneapolis: Fortress, 2009), *Creaturely Theology*, edited with David Clough (London: SCM, 2009), *Religion and Ecology in the Public Sphere*, ed. with Heinrich Bedford-Strohm (New York: Continuum, 2011), *Animals as Religious Subjects*, edited with Rebecca Artinian Kaiser and David Clough (London: T & T Clark/Bloomsbury, 2013), *The Wisdom of the Liminal: Evolution and Other Animals in Human Becoming* (Grand Rapids: Eerdmans, 2014), *ReImaging the Divine Image* (Kitchener, ON: Pandora, 2014), *Technofutures, Nature and the Sacred*, edited with Bronislaw Szerszynski and Sigurd Bergmann (Farnham, UK: Ashgate, 2015), *Ecology in Jürgen Moltmann's Theology*, 2nd ed. (Eugene, OR: Wipf and Stock, 2016), and *Religion in the Anthropocene*, edited with Sigurd Bergmann and Markus Vogt (Eugene, OR: Cascade, 2017).

Heather Eaton

Heather Eaton is a professor at Saint Paul University, Ottawa. She holds an interdisciplinary PhD in ecology, feminism, and theology. She is engaged in religious responses to the ecological crisis: the relationship among ecological, feminist, and liberation theologies, and connections between religion and science. She is committed to interreligious responses to ecological crisis. Her authored and edited books include *Advancing Nonviolence and Social Transformation: New Perspectives on Nonviolent Theories*, with Lauren Levesque (Sheffield: Equinox, 2016); *The Intellectual Journey of Thomas Berry: Imagining the Earth Community* (Lanham, MD: Lexington, 2014); *Ecological Awareness: Exploring Religion, Ethics and Aesthetics*, with Sigurd Bergmann (Münster: LIT Verlag, 2011); *Introducing Ecofeminist Theologies* (London: T & T Clark, 2005), and *Ecofeminism and Globalization: Exploring Religion, Culture, Context* (Lanham, MD: Rowman and Littlefield, 2003), with Lois Ann Lorentzen, plus dozens of academic articles. She is on the board of the journal *Worldviews: Global Religions, Environment and Culture*, the steering committee of the Religion and Ecology session of the

American Academy of Religion, and is past president of the Canadian Theological Society.

Matthew Eaton

Matthew Eaton is a postdoctoral teaching fellow at Fordham University in New York, where he teaches theological ethics and religion and science. He received his PhD in 2017 from the University of St. Michael's College in Toronto. His dissertation is entitled "Enfleshing Cosmos and Earth: An Ecological Theology of Divine Incarnation." He has published several articles on posthumanist ecologies and environmental ethics from various perspectives in the continental tradition, focusing on the work of Emmanuel Levinas and Maurice Merleau Ponty. He is slightly obsessive about cats, and his feline companion animal, Fargo, rules his life and home.

Timothy Harvie

Timothy Harvie is Associate Professor of Philosophy and Ethics at St. Mary's University in Calgary, Canada. He received his PhD at the University of Aberdeen and a postdoctoral licence at the University of Wales, Lampeter. He is the author of *Jürgen Moltmann's Ethics of Hope: Eschatological Possibilities For Moral Action* (Aldershot, UK: Ashgate, 2009) and has published articles on eschatology, Thomas Aquinas, economics, and ethics. His research interests include theological anthropologies, the political economy, continental philosophy, environmental theology, and animal studies.

Laura Hobgood

Laura Hobgood is Professor and Paden Chair in Religion and Environmental Studies at Southwestern University (Georgetown, Texas), where she has served on the faculty since 1998. Dr. Hobgood has chaired both the religion department and the environmental studies program. She teaches in the areas of religion and ecology, animals in religion, the history of Christianity, environmental studies, animal ethics, and ecofeminism. Her most recent books are *A Dog's History of the World: Canines and the Domestication of Humans* (Waco, TX: Baylor University Press, 2014), *The Friends We Keep: Unleashing Christianity's Compassion for Animals* (Waco, TX: Baylor University Press, 2010), and *Holy Dogs and Asses: Animals in the Christian Tradition* (Champaign, IL: University of Illinois Press, 2008). She also served

Cristina D. Vanin

Cristina Vanin, PhD, is associate professor of theology in the Department of Religious Studies, Associate Dean, and Director of the Master of Catholic Thought program at St. Jerome's University, Waterloo, Ontario. Her areas of research include ecotheology, feminist theologies, and Christian ethics. Dr. Vanin's teaching and research draws on the thought of the cultural historian Thomas Berry, as well as the work of Canadian philosopher and theologian Bernard Lonergan. Her most recent publication is: "Understanding the Universe as Sacred," in *The Intellectual Journey of Thomas Berry: Imagining the Earth Community*, edited by Heather Eaton (Lanham, MD: Lexington, 2014).

Mark Wallace

Mark Wallace holds a PhD from the University of Chicago, and serves as a professor in the Department of Religion and member of the Interpretation Theory Committee and the Environmental Studies Committee at Swarthmore College. His teaching and research interests focus on the intersections between Christian theology, critical theory, environmental studies, and postmodernism. His authored publications include, besides dozens of articles, *Green Christianity* (Minneapolis: Fortress, 2010), *Finding God in the Singing River: Christianity, Spirit, Nature* (Minneapolis: Fortress, 2005), *Fragments of the Spirit: Nature, Violence, and the Renewal of Creation* (New York: Continuum, 1996; Harrisburg, PA: Trinity, 2002), *The Second Naïveté: Barth, Ricoeur, and the New Yale Theology* (Macon, GA: Mercer University Press, 1990, 1995). He has also served as editor of Paul Ricoeur's *Figuring the Sacred: Religion, Narrative, and Imagination* (Minneapolis: Fortress, 1995), and coeditor of *Curing Violence: Essays on René Girard* (Sonoma, CA: Polebridge, 1994). He is a member of the Constructive Theology Workgroup, active in the educational justice movement in the city of Chester, and recently received an Andrew W. Mellon New Directions Fellowship for a research sabbatical in Costa Rica.

Preface

CHRISTOPHER CARTER

Growing up in a black church I was surrounded by a community of church Mothers, and there are certain sayings among the Mothers that a child would hear from time to time. As a young boy, sayings such as "Weeping may endure for the night, but joy comes in the morning," "We have to make a way out of no way," and "This too shall pass" were etched into my consciousness. Among my favorites is a paraphrase from the beginning of Psalm 8, "out of the mouths of babes." This saying signaled that a young child, in a moment of sheer childlike honesty, might offer insight into a problem that the adults in the room had seemingly overlooked. In these moments, the "simple" musings of a child were given adult-like credibility. Wisdom was thus no longer confined to the margins of community meaning-making; instead, the kernel of truth was placed in the fertile ground of elder wisdom and allowed to grow.

As a curious, nosy, and (at least in my mind) smart child, I loved hearing "out of the mouths of babes" in reference to something that I said. I longed to be a part of the community that was *allowed* to fashion meaning out of everyday existence. Similarly, *Encountering Earth* asks us to consider why we haven't given voice to the other-than-human materiality that helps shape our ideas about God-talk. Indeed, despite knowing what it feels like to be silenced, most theologians either rationalize the silencing of the more-than-human world or understand human reason to be the hermeneutical key necessary to give voice to the "others." This volume is an attempt to think theologically in community with the more-than-human-world, and in doing so, it presumes that the non-humans who occupy the world are active agents, expressing themselves within the broader Earth community.

Our encounters with the world give shape to theological ideas. Our encounters shape both our reason, how we make sense of the world, and our being, how we feel about the world we inhabit. To be sure, the vestiges of Enlightenment thinking where a small group of white men convinced much of humanity that reason should take priority over affect, are still present in the modern world. Unsurprisingly then, we have muted our theological reflection upon our encounters with Earth. More specifically, such judgments have undervalued being as such, by rationally interpreting existence through the lens of human dominance of other species and white male heterosexual dominance over all human beings. In this way, *Encountering Earth* is a radical liberative invitation to encounter others in their fullness while simultaneously experiencing the encounter as our full selves, beyond the narrow definition of humanity framed during the Enlightenment. Similar to others in this volume, my life and my understanding of the Divine has shifted as a result of my encounter with Sampson, my Chocolate Labrador.

I didn't want another dog. My wife called me and asked if I would come to her office to visit a dog she found while running errands for her boss. Of course, I agreed to meet the dog, I even consented that we could foster the dog until we located his owners, but I told her that three dogs were enough. To be sure, we had the space for another dog. We lived in a large four-bedroom house on an acre of land, but that was beside the point. I knew another dog meant more work for me, more cleaning, more training, more walking, a whole lot of more. I didn't want another dog. When I arrived at her office, my wife walked me back to the office where she was keeping the dog and said, "this is Sampson." "She's named him," I thought to myself; I could see where this was going. So, I took Sampson home. He met our other dogs, Zeus, Zeke, and Cookie, and got along fine with them. I was surprised at how calm he was. It was almost as if he knew that I was looking for a reason to take him to the shelter. It turns out that Sampson was perhaps the rare Chocolate Labrador that was, in fact, calm by nature. He loved being in community, he would follow the other dogs around, he would lay on the floor in front of me while I watched TV, he didn't want to distract you, but he wanted you to know he was present. As the days passed I could hardly tell that we added a fourth dog to our family and I accepted that Sampson was here to stay.

Two years later, my wife, Zeus, Sampson, and I were driving across the country to Claremont, California, where I was to attend graduate school. Given that we were going to downsize and live in an apartment, we decided that Cookie and Zeke, the most rambunctious of the pack, would be better suited living with friends in Michigan who owned houses. Sampson took to California quite easily. He loved the sun and the beach. His laid-back personality was a great fit for the relaxed "bro" vibe of Southern California. Unfortunately, the relaxed nature of our household was upended when my wife and I experienced the seemingly predictable graduate school marriage crisis. Our relationship struggles eventually resulted in a four-month separation that was particularly hard on everyone. After my wife moved out of the apartment, I sunk into a deep depression. During the first few weeks, I spent most of my days on the couch. I wasn't the best companion during this time and Sampson had every right to be annoyed at my inability to maintain a healthy and active relationship with him and his brother. Sampson stepped into this joyless void. He was exceptionally affectionate during my depression, often laying his head on me so that I could feel his presence. Again, he wanted me to know that he was present, and in these instances, he must have thought that I needed to feel his presence as well. He didn't do anything particularly novel other than ensuring that I was never alone, which in hindsight, was exactly what I needed. A few years later, I would return the favor.

After finishing my doctoral studies, I accepted a post-doctoral position at the University of San Diego. Several months earlier we lost Zeus to a soft-tissue sarcoma and now Sampson was our only companion. Since my wife was now in veterinary school, we were going to have to become a two-house family. We decided that Sampson would live with me in San Diego because I would have a more flexible schedule and would be able to spend more time with him now that he was going to be alone during the day. However, this would be my first time teaching an undergraduate class and I quickly realized that I didn't know what I was doing. I spent more time on campus preparing for class than I anticipated which resulted in Sampson being home alone much more than either of us liked. I went to several "new faculty" events that occupied my time as well. Sampson was home by himself much more than he should have been, but I believed I was doing what all great professionals do, I was proving my worth to the university.

One day we were preparing to visit my wife for the weekend and I noticed he was moving slower than usual. I didn't think anything of it; he was getting older after all. When we arrived at her townhome Sampson was breathing heavy, but I thought it was due to the heat because it was a particularly hot fall that year. When my wife saw him she immediately knew something was wrong and we rushed to the hospital. Sampson had developed a cancerous tumor on his heart, a hemagiosarcoma, which allowed fluid to build up around his heart and limit its ability to function. Sampson was dying. After the doctors drained the fluid that had built up around his heart he seemed like he was back to his old self. I didn't know how much time he had left, but I knew that I needed to spend more time with him to ensure that he was doing ok.

I changed my work habits and started working from home as much as I could. I stopped attending so many university events and reduced my presence around campus. At first, changing my schedule was easy because it seemed like the right thing to do. But eventually, people start wondering why you are not around as much as you were. Perhaps, most importantly, I began feeling like I was missing something by not being on campus or attending as many events. I started to resent having to come home every few hours to walk Sampson and I was frustrated that he seemingly had so much control over my schedule. Sampson had always done such an excellent job of blending in and being low maintenance that I didn't know what to do when he required so much additional attention. Moreover, I loved Sam and I was disappointed in myself for feeling any resentment toward him; it's not like he wanted to have cancer. And yet, it seemed that the various feelings I had toward him never dissuaded his opinion of me. I was a part of his community, and I belonged to him as much as he belonged to me.

Five months after his first fluid drain Sampson had a second procedure. After this hospital visit we knew that we only had a short time left with him, and it turned out that we just had two days. The night he passed away we went for his usual evening walk and he smelled all his favorite smells. When we got back to the apartment he wasn't able to walk up the two steps required to get inside. I knew it was time. I carried him inside, laid him on the couch and we sat and watched TV for a few hours. We both fell asleep on each other, comforting each other in a way similar to how he comforted me during my depression. My wife and I drove him to the hospital in the morning to say goodbye; I have never wept so hard in all my life.

The following day my wife returned to our townhome in Claremont while I stayed in San Diego. I woke up around 6:30 AM, the usual time Sampson would wake me up for our walk. I got dressed and went on "our" walk by myself. I walked every morning and every evening for several days and I as I walked I tried to process how I felt about Sampson's death. I felt a piercing sense of guilt about the last few months of Sampson's life. I felt ashamed that I had been so selfish; the manner I had lived these past few months was inconsistent with the ethical values that I claim to uphold. In this way, my sense of guilt was more than just feeling distressed that Sampson was dead: I felt as though I had sinned, against Sampson and God.

As a liberation ethicist, I understand sin to be a communal, rather than individual concept. I agree with James Cone that "to be in sin is to deny the values that make the community what it is. It is living according to one's private interests and not according to the goals of the community.

It is believing that one can live independently of the source that is responsible for the community's existence."[1] I was a part of Sampson's community and the last few months of his life I wrestled with the moral obligations that my membership required. Growing up as a poor black kid and now as a black academic, I know what it feels like to be an invisible member of multiple communities—to be present and not be seen, to be asked to speak and know that no one is listening. As such, my liberatory anti-oppression ethic is rooted in making visible what we would rather not see, giving voice to those we'd rather not listen to, and welcoming people into the beloved community whom the dominant culture has told don't belong.

In his life, Sampson showed me that expanding our notion of community is rarely as challenging as we make it out to be. In his death, he revealed how easy it could be for anyone to fall prey to human selfishness and shrink our communities, and therefore reduce our moral obligations toward others. In short, through my encounter with Sampson, sin is no longer an abstract theoretical concept rooted in Euro-Christian norms of moral impurity. My theological description of what sin is and how it (mis)-shapes communities can only make sense in light of my encounter. Those who dedicate themselves to the pursuit of love and justice rightly recognize that we will be confronted by forces that seek to stand against our goal of fashioning a liberatory anti-oppressive community. History shows us that if we are persistent in our struggle we may be invited to the negotiating table of those who hold some power within the dominant culture. It is at the negotiating table where we will have to discern how much and for whom are we willing to compromise our theo-ethical commitment to those on the margins of our community. Sampson showed me how easy it is to settle for an alleged permanent seat at the table, to *settle* for equality with, rather than liberation from, the oppressive structures we claim to fight against because we have bought into the idea that to include all of the marginalized at the table would be too inconvenient.

Abstract theological ideas such as sin, faith, and hope reflect the moral strivings we project onto Divine character when they are explicated in ways that dismiss our encounters with others, including the Earth. These abstractions universalize the affective dimension of moral action, and in doing so causes us to either dismiss any feelings that contradict what we are supposed to feel or rationalize our desire to avoid feeling compassion for marginalized groups. By theologically attending to our encounters with Earth we can begin to dismantle the logic that attempts to separate these encounters from

1. James H. Cone, *A Black Theology of Liberation*, 40th Anniversary Edition (Maryknoll, N.Y: Orbis Books, 2010), 110.

our understanding of the Divine. For me, Sampson is as much of a prophet as any within the Christian tradition. Within the Christian tradition our communities are the source of our life, time and time again, he pulled me back to that source. As such, my notion of the Divine pull toward a beloved community will forever be interpreted in light of my being a part of Sampson's community. My hope is that in reading this text you too can reflect on your encounters with Earth and that our world may be all the better for it.

Introduction

—TIMOTHY HARVIE, MATTHEW EATON, AND TREVOR GEORGE HUNSBERGER BECHTEL—

> Yahweh then gave the donkey the power to talk, and she said to Balaam, "What harm have I done you, for you to strike me three times like this?" Balaam answered the donkey, "Because you have been making a fool of me! If I had been carrying a sword, I should have killed you by now." The donkey said to Balaam, "Am I not your donkey, and have I not been your mount all your life? Have I ever behaved like this with you before?" "No," he replied. Yahweh then opened Balaam's eyes and he saw the angel of Yahweh standing in the road with a drawn sword in his hand; and he bowed his head and threw himself on his face. And the angel of Yahweh said to him, "Why did you strike your donkey three times like that? I myself had come to bar your way; while I am here your road is blocked. The donkey saw me and turned aside because of me three times. You are lucky she did turn aside, or I should have killed you by now, though I would have spared her."
>
> — Num 22:28–33

Although Balaam's donkey does not have a name—unlike many of the creatures you will encounter in this volume—she is an individual and her particularity is both recognizable by and decisive for Balaam. Likewise, her particularity, expressed here in her concern for her own well-being, is recognizable by and decisive for God—"Why did you strike your donkey three times like that?" Balaam's donkey, however, has been neither recognizable nor decisive for the vast majority of humans who have reflected on this more-than-human story with theological, biblical, or philosophical lenses. With Balaam, we often refuse to recognize and respond to others. We refuse to be called into question by the more-than-human world and reject the consequent call for conversion that inevitably follows. Such calls are indicative of voices not only creaturely, but simultaneously divine.[1] Such refusals do not acknowledge other-than-human bodies and expression as necessary for theological reflection and—perhaps surprisingly given how important the particular is in Christian reflection—actively silences more-than-human animals, landscapes, or geographies when, in defiance of the ban placed upon them, they do speak. Nevertheless, Balaam eventually encounters the donkey's voice and recognizes the revelation implicit within it. Likewise, some humans do encounter the more-than-human world and respond favorably to the divine expression of the creatures therein. This book tells the story of some of those encounters and offers the theological reflections of the humans involved in them.

Historically, theologies of creation and animal theologies have occupied a minimal place in the great dogmatic or systematic works of theology. Theologies of creation have labored to give an account of creation "in the beginning" and systematic theology has generally engaged creation as it explores the nature and place of sin (hamartiology), or has located creation as a subsection or sub-discipline under broader themes of theological anthropology or the doctrine of redemption.[2] The natural world, as it speaks for itself prior to human commentary, has been neglected as a resource for theological construction. Only with the realization of an emerging ecological crisis have politically engaged theologies begun to take seriously the role of Earth and its varied inhabitants for its own sake and the fate of humanity.[3]

1. The models for contemplating divine/creaturely expression are plural and take various forms in the essays in this book. As such, we refrain from delving into this further as we introduce a plurality of voices speaking to this common theme. It is sufficient for now to recognize that the narrative of Balaam's donkey insists that the creature speaks on its own behalf and for its own concern and that such an expression is tied to divine expression and revelation.

2. Gerhard von Rad, for example, strongly subjugated creation to redemption in an essay first published in German in 1936 that had broad influence over a generation of scholars, "The Theological Problem of the Old Testament Doctrine of Creation," 142. Von Rad later retracted these ideas in *Wisdom in Israel*. Ellen Davis, in *Scripture, Culture, and Agriculture* and Norman Habel in the Earth Story series are examples of a different biblical approach to creation.

3. We cannot, of course, ignore the response of Pierre Teilhard de Chardin to Darwinian evolution, though Teilhard does not, strictly speaking, write out of the same contemporary concern that today's eco- and animal theologians write. Following the

This is particularly true of political theologies attempting to reflect critically upon the degradation of nature as a result of human activity.[4] Peter Scott, for example, has written a "political theology of nature" that attends "to the interaction between un/natural humanity and socialised nature," but does so as "an exercise in theological anthropology" thus maintaining the centrality of the human.[5] Alternatively, Michael Northcott has sought to develop a "political theology of climate change" that recognizes the inherently political nature of investigations into climate and the growing disconnect between humans and the other-than-human world.[6] Much of this literature grapples with the theological and religious challenges of living in what is called the Anthropocene.[7] Until recently, then, modern theological contemplation of the more-than-human world for its own sake has been something of an anomaly. The importance of such contributions cannot be overstated, but even these do not often attend to the particular voices of creatures who inhabit Earth alongside humans. This is the goal of emerging theologies focused on specific other-than-human animals.[8] Here the interrelationships between human and other-than-human animals that have been marginalized in traditional theology are now beginning to receive sustained attention.[9] Animal theology, and animal studies more broadly, are rapidly

---

early work of Teilhard de Chardin, the pioneering work of Joseph Sittler, and a seminal essay Lynne White, political theologies of Earth began to take firm shape in the 1970s and 1980s. Prominent examples include: Santmire, *Brother Earth*; Cobb, *Is It Too Late?*; Derr, *Ecology and Human Liberation*; Ruether, *New Woman, New Earth*; Moltmann, *God in Creation*; McFague, *Models of God*; McDaniel, *Of God and Pelicans*.

4. Johnson-DeBaufre et al., eds., *Common Goods: Economy, Ecology, and Political Theology*.

5. Scott, *A Political Theology of Nature*, 4–5.

6. Northcott, *A Political Theology of Climate Change*, 15–21; and *Place, Ecology and the Sacred*, 1–3.

7. Bonneuil and Fressoz, *The Shock of the Anthropocene*; Deane-Drummond et al., eds., *Religion in the Anthropocene*.

8. A pioneer in animal theology is Andrew Linzey, who began addressing the issue of animal theology in the 1970s. See especially *Animal Rights; Christianity and the Rights of Animals; Animal Theology*; *Animal Gospel*; *Creatures of the Same God*; and *Why Animal Suffering Matters*. See also McDaniel, *Of God and Pelicans*; Pinches and McDaniel, eds., *With Roots and Wings; Good News for Animals*; MacKinnon and McIntyre, eds., *Readings in Ecology and Feminist Theology;* and Webb, *On God and Dogs*. Since these writings, the field has seen significant development that we cannot comprehensively list here.

9. Beyond Linzey and others, the work of David Clough stands out as pushing animal theology in a promising direction. Clough, *On Animals Volume 1*. Other collected volumes demonstrate just how many people are now working on this topic. See for example, Waldau and Patton, eds., *A Communion of Subjects*; Deane-Drummond et al, eds., *Animals as Religious Subjects*; Moore, ed., *Divinanimality*. For a historical

growing fields that embrace a multidisciplinary approach to the question of the other animal, even if such studies are often abstract meditations that do not explicitly locate creatures themselves as the locus of revelation and theological ethics. This volume seeks to bring both of these emphases into focus, explicitly exploring how the more-than-human world in all its variegated forms express to humanity and inspire religious thought by virtue of their own unique, revelatory voices.

Philosophical and theological accounts arising from the pressing question of how we relate to and hear the expression of non-humans, animal or otherwise, form the background of this volume.[10] While there are differing approaches to the particularity of the non-human other, this volume builds upon the idea of intersubjectivity to explore how concrete encounters with the more-than-human world manifest the embodied affect that arises from our own animality and finally erupts in religious reflection.[11] Essential to these affective encounters is the affirmation of the relational capacities of all things in their own multiform subjectivities. These relational capacities generate creative opportunities in the liminal spaces between beings and the manifest divinity in those experiences. In this, the authors of this volume reflect on the manifold ways in which divinity and materiality are entangled and how these are revealed in concrete encounters.

This approach differs from how classical theology has treated the non-human world. We recognize that the Book of Nature, along with the Book of Scripture, has informed Christian theology throughout the ages. For example, the Bestiary tradition reflected on the natural history of creatures and sought to glean learnings for humans from nature. The Book of Nature, however, was always understood as an expression discerned by means of human rationality and interpretation, and for human understanding and living. Humanity has thus been understood as the hermeneutical key required to rationally discern the divine meaning assigned to the natural order. This volume is not a work of natural theology. This work presumes that the more-than-human world is other than a passive object to be interpreted, and instead an active plurality of voices that reaches out and expresses itself within the boarder Earth community.[12] This expression shapes the very pos-

exploration of Christianity and animals, see Hobgood-Oster, *Holy Dogs and Asses*.

10. See Calarco, *Zoographies*; Abram, *Becoming Animal*; and Gross, *The Question of the Animal and Religion*.

11. See Schaefer, *Religious Affects*.

12. This type of embodied, non-logocentric speech is best understood in the framework of Maurice Merleau-Ponty's philosophy of perception. See especially Merleau-Ponty, *Phenomenology of Perception* and *The Visible and the Invisible*, as well as David Abram's *The Spell of the Sensuous* and *Becoming Animal*, a specifically

sibility of human thought, speech, and rationality. The voice of the more-than-human world then, in a real sense, makes theology possible and the essays in this volume suggest that what we say theologically represents not simply ideas of our own making and subsequently superimposed onto the natural world through our own discovery. We suggest that theological ideas are formed in response to the expression of what is fundamentally otherwise than human. In such a relationship with the more-than-human world, the human is initially passive—i.e., the divine entangled in the material reveals itself beyond human rationality and conditions its very possibility. We recognize that we are not constructing theology out of nothing but our rational minds; we are actively responding to, or remembering, a revelation that erupts within a more-than-human world to which we were once passively exposed. These encounters with Earth are what makes thought, speech, rationality, and ultimately theology, possible. As such, the theologies in this volume are a recounting and remembering of what we have heard and felt in our face-to-face encounters with Earth—encounters we understand to be inseparable from an encounter with divinity. Theology then is inherently grounded in these intersubjective, material encounters, rooted in and conditioned by a more-than-human world.

Just as Jacques Derrida is able to build a philosophy of the more-than-human from an encounter with a particular feline, this volume is grounded not in metaphysics nor an ontotheological framework but proceeds by way of a hermeneutical appropriation of revelatory experience itself.[13] By beginning with the affective encounter, rather than a theoretical framework, the essays in this volume reflect what Matthew Calarco argues when he states, "What thought will encounter once reliance upon these categories is surrendered cannot be known in advance; however, it is certain that any genuine encounter with what we call animals will occur only from within the space of this surrender . . . in the clearing of the space for the event of what we call animals."[14] Similarly, Donna Haraway has argued that "We also live with each other in the flesh in ways not exhausted by our ideologies. Stories are much bigger than ideologies. In that is our hope."[15] The rawness of these

eco-phenomenology and philosophy of non-anthropocentric expression rooted in Merleau-Ponty's philosophy.

13. Derrida, "The Animal That Therefore I Am." We thus do not claim to represent our encounters with absolute accuracy or to capture the voice of Earth. Theologies of nature such as these ultimately are our own imperfect expressions of remembrance of the encounter. Yet, we do insist that the ground of such theology lies in a revelation beyond human thought, which we now wrestle to understand.

14. Calarco, *Zoographies*, 4.

15. Haraway, *The Companion Species Manifesto*, 17. See also, Haraway, *When Species*

embodied encounters with the more-than-human world, and the narratives that arise from them, are loci of theological construction where the ways in which animal, plant, element, and cosmos are entangled with the divine and are reflected upon in myriad ways: sacramentally, animistically, philosophically, scientifically, and cosmically.

Therefore, the essays in this book place such narratives of encounter front and center. They are the ground upon which theology is built and made possible. They begin with descriptions of actual encounters and attempt to understand the ways in which these experiences have revealed God prior to our inherited ideas about what divinity entails. The essays are written by those who theologize, broadly speaking, within the Christian tradition, though in recognition that the tradition may be questioned and critically engaged based on our encounters with Earth. Theology thus arises from the encounter, and not from some rational argument that convinced us that the otherwise-than-human should be theologically considered. As novelist J. M. Coetzee writes:

> We (participants in this dialogue) are where we are today not because once upon a time we read a book that convinced us that there was a flaw in the thinking underlying the way we, collectively, treat nonhuman animals, but because in each of us there took place something like a conversion experience, which, being educated people who place a premium on rationality, we then proceeded to seek backing for in the writings of thinkers and philosophers. Our conversion experience as often as not centered on some other mute appeal of the kind that Levinas calls the look, in which the existential autonomy of the Other became irrefutable—irrefutable by any means, including rational argument.[16]

Coetzee speaks of the animal other—and if we were to read him further we would see he has much more than animality in mind—only after being addressed by its alterity. For Coetzee, the more-than-human voice addresses and converts the human through something of a paradox, an expression made by means of a "mute appeal," an expression beyond a logocentric understanding of language.[17] This is the spirit behind the essays in this volume.

Meet.

16. Coetzee, "Notes on Issues Raised by Matthew Calarco," 89.

17. While all thought might take place within an inherited language that betrays the experience reflected upon, this does not refute the idea that language is rooted in and made possible by experience deeper than constructed rationality, nor does it eschew the idea that experience shapes the possibility and contours of thinking within an inherited language or challenges previously accepted ideas altogether.

Each author has encountered something of Earth and has been converted by some face-to-face experience, and thus writes in remembrance of how Earth expresses a divine voice that grounds and shapes religious thought.

The chapters are arranged geographically and in four sections. We begin within the human family and home and move outward to farm and lab, wilderness and wild, and finally cosmos and Earth.[18] In organizing the essays in this way, we invite the reader to journey with the authors from those environs that seem most familiar, even mundane, and explore the ever-increasing breadth of potential for theological, moral, and spiritual transformation as the emergent possibilities of encounter with the transcendent occur in and through an encounter with various faces of Earth and beyond. You can view pictures of the cats and dogs, snakes and landscapes, encountered in this book at https://www.facebook.com/encounteringearth/.

The first section of essays examines encounters under the rubric of "Family and Home." Trevor Bechtel and John Berkman each reflect upon loss. Bechtel's chapter recounts the life and death of Tiamat the cat, who entered his home and transformed it through her independence and those moments she chose to embrace her human companions. Bechtel attempts to understand this transformation by analogy to the inner-relations of God via Eugene Rogers and inside the larger domesticated context needed to support or contain Tiamat's life via the death of an unnamed chicken. The death of Tiamat and the transposition of this being from a friend who is known to something else entirely receive theological exploration here. Analogously, Berkman investigates his own experience in welcoming a Retriever-Great Dane cross named Max into the family only to surrender that relationship after Max struggles to interact with other humans—the lamentable fruit of previous neglect and abuse. This experience is viewed in light of Job 39 where Job is challenged to see living creatures in light of their fullness in God in the midst of his own suffering. Max is emblematic of the brokenness of the world in its current state. Matthew Eaton continues exploring relations with companion animals in recollecting a chance encounter with a kitten who demands recognition and acknowledgement. Eaton notes the communicative capacity of caged animals and the possibility of acknowledging affective appeals for creaturely flourishing, regardless of the linguistic distance between species. It is in the exposure to and of the alterity of a small cat that Emmanuel Levinas's infinite moral responsibility becomes manifest. Through this lens, Eaton recounts how he is able to see the divine face of the other in the eyes of this cat and the curve of his claw. Timothy Harvie's chapter reflects upon a dog who had experienced neglect in his

18. See Urbanik, *Placing Animals*.

house of origin and the challenges that come with inexperience and the frustration of anxiety. However, here the outcome differs when the young dog initiates a peaceful ritual of touch, which becomes an opportunity for the inhabitation of eschatological life. Embodied touch becomes the entrée of God into the space between these two and redeems them both. Grace Kao's essay reflects upon the intersections of race, family, and interspecies relationship in recounting her encounter with a childhood cat. Kao's relationship with Morris engages the complex social realities of being raced as a Taiwanese-American and the interstitial cultural spaces that bear the tension of membership in multiple worlds. A theological understanding of the notion of friendship is developed in critical dialogue with the reality of domesticity in a trans-cultural home.

The second major section captures the domesticated locales of the "Farm and Lab." Celia Deane-Drummond shares her childhood experiences with ponies and horses in order to consider the implications of human-equine relations and expand upon these in light of ethnographic studies of the Oromo people in Ethiopia and the wider horn of Africa. These examples illustrate what she terms "inter-morality" to express complex interspecies ethical relations. Here wisdom is the primary category deployed, which is able to bypass the complex issues pertaining to a theory of mind and instead explore a theory of perception rooted in human/horse interactions grounded in our communal evolution. Deane-Drummond's essay continues to develop her explorations in theological anthropology as she offers further ways in which human and divine wisdom is intimately tied to animal relations. Laura Hobgood's chapter initiates a new avenue of exploration as she ponders the significance of a non-domesticated species of animals, namely rats. In contemplating the lives of laboratory rats on her campus in light of a childhood experience of caring for a mother rat and her offspring, Hobgood analyzes these realities in light of Christian reflections on sacrifice in theology and spiritual practice. The harrowing realities of the sacrificial lives of many animals at human hands is exposed in light of a criticism of the elevation of theologies that state that Jesus was sacrificed to elevate only the human. Hobgood calls into question the culture of animal sacrifice and the arbitrary elevation of the human by challenging these narratives and proposing resources where a Sabbath peace is available through the life of Jesus who, in the gospel of Mark, is described as being with the wild animals. Finally, Abigail Lofte recalls being raised on a farm and the manner in which land was demarcated in domestic and undomestic spheres. In dialogue with Edward Schillebeeckx, interpreted through the ecologically centered framework of Thomas Berry, Lofte sets her reflections on the role of the farm in the context of an earth-oriented, sacramental Christology where the resurrection of Christ provides the impetus for Earth justice. Such justice arises from the combination of the gospel of tradition, emphasizing the liberative focus of Jesus' resurrection, with Schillebeeckx's notion of the "fifth gospel," which suggests that Christianity is incomplete without the narrative of a person's own experience in the world. For Lofte, the fifth gospel was written on her family farm, where the creative tension of animal, plant, and element living together shapes her understanding of theology.

The third section is yet broader in its geographical scope and inhabits the untamed places of nature that we name "Wilderness and Wild." Kimberly Carfore begins this section by recalling an encounter with a venomous Copperhead snake in the Blue Ridge Mountains of Appalachia, whom she is forced to kill. Carfore writes of the liminal places between finitude and infinity as she seeks to understand the symbol of the snake's lifeless body in her experience. In a Derridean analysis, she deconstructs the notion of God and religion in order to note the potential of salvific justice in the slithering other. Recognizing the ambiguity of human existence, Carfore wrestles with divine, creaturely speech and the tension arising when humans recognize their deep desire for peace despite the inescapable nature of violence. Lisa Dahill moves beyond animality and reworks the Christian role of the "Book of Nature" as revelatory while kayaking upon rivers teeming with life. These waters express the vibrancy of the ancient prescription that the rite of baptism unfold within "living waters." This is contrasted with the modern human compulsion to dam waters and thus transform them from living into deadly waters. This is mirrored in the stagnant baptismal font in many indoor parishes. In re-entering living and wild waters, Dahill reflects upon the potentialities for connections between animist accounts of nature and Christian spiritual practice. Dahill's is an elemental theology, exploring the feminist deconstruction of Luce Irigaray and Ellen Armour, the ecophenomenology of David Abram, and the ethics of Dietrich Bonhoeffer for a theology that recognizes that intersubjective expression exists beyond animality, in the watery veins that support earthly existence. In the next chapter, Nathan Kowalsky exposes the reader to a harsh dimension to the relationship between humanity and wildlife by encountering the hunt. In the taking of life of a deer, Kowalsky reflects on his own mortality, the hiddenness and wildness of God, and the inescapable realities of violence and survival. Kowalsky's encounter is one of acknowledged violence, but nonetheless one that still recognizes the concrete specificity of the prey through his role as predator. In dialogue with the continental philosophical tradition, and especially the thought of José Ortega y Gasset, the hunt is not entertainment or sport, but a way of being in the wilderness that still recognizes divinity in the challenging dynamic of predation. Cristina Vanin's essay brings the reader back to cultivated spaces in a retreat setting that draws out the implications for retreat as a form of spiritual practice. She is attentive to the work of Thomas Berry, Lyanda Haupt, and contemporary magisterial reflection on creation against the backdrop of the larger scientific picture of the universe and makes important connections between particular retreat spaces and this larger story.

The final section synthesizes the whole in contemplating "Cosmos and Earth" as loci of theological construction. Jame Schaefer reflects upon the manner in which a life of advocacy and intellectual, scientific inquiry has been shaped by the land and natural environment surrounding her. This journey was initiated through protest and advocacy over the construction of an interstate, which would transform the important local environment into a freeway for cars. In locating a theological journey in such an origin, Schaefer's essay reflects a biography of theological construction impacted by a deep encounter with Earth. Mark Wallace continues his legacy of encountering God as avian mystery and majesty. In furthering a development of his understanding of Christian animism, Wallace remembers a chance meeting with the "Lord God bird" in the Crum Woods, on the edge of the Swarthmore College campus. In dialogue with both the theological and non-theological study of religion, this essay challenges the view that Christianity is an unearthly religion with little to say about everyday life in the natural world. Wallace rebels against the war some Christians wage against the flesh in the denigration of the body, arguing that an understanding of Christianity as an unearthly religion that is hostile to the creaturely world, while historically accurate, misses the supreme value biblical religion assigns to all of the denizens of God's creation, human and more-than-human alike. Wallace's approach to Christian animism unfolds in dialogue with his own encounters with divine avian Earth, which extends to all flesh. Heather Eaton concludes the volume with the most expansive essay of the collection. Eaton takes us beyond Earth, recognizing that while this planet is our immediate home, we exist within a much wider cosmic community—all of which is entangled with divinity. Through a reflection on the wonders of stargazing, Eaton leads us into the curiosity required to engage the dynamics of the universe and get to the heart of some of the most significant and classic human quests: Where we are? Who we are? How are we to live? Her wrestling with these questions point her to argue, following Pierre Teilhard de Chardin, that life in cosmos and Earth takes place within a fundamentally *divine milieu*. In the quests to answer these questions, Eaton concludes with a reflection on ethics and the way forward in an age of ecological crisis. She, like Thomas Berry before her, calls for a transformed cultural orientation that insists on embracing our relationships with our more-than-human neighbors, a robust scientific view of the world, and religious claims that function in the interests of a vital biosphere. If we can shift our culture toward such ends, she insists that there may be hope in the midst of crisis.

These essays represent a novel approach to theological thinking in the academic world. Yet, we recognize that such ways of thinking are by no means novel in the wider world, where encounters with Earth have shaped religious thought for ages. As such, these essays are a quantitatively insignificant representation of the concrete encounters shared by many throughout the world and throughout time. We hope that reading about such encounters bring to mind the ubiquitous and particular encounters that have already occurred in the lives of the reader. Such encounters will ideally change the way we think not only about God and religion, but about the world in which we live and move and have our being. The cultural transformation envisioned in such a change of perspective echoes the growing recognition within the theological and non-theological studies of religion that if our species is to live peacefully on Earth, insofar as this is possible, religion must embrace its encounters with Earth and the manifold ways in which divinity might speak, and even call us into question, in these meetings.

Bibliography

Abram, David. *Becoming Animal: An Earthly Comsology*. New York: Pantheon, 2010.

———. *The Spell of the Sensuous: Perception and Language in a More-Than-Human World*. New York: Vintage, 2003.

Bonneuil, Christophe, and Jean-Baptiste Fressoz. *The Shock of the Anthropocene*. Translated by David Fernbach. New York: Verso, 2016.

Calarco, Matthew. *Zoographies: The Question of the Animal from Heidegger to Derrida*. New York: Columbia University Press, 2008.

Clough, David L. *On Animals Volume 1: Systematic Theology*. New York: Bloomsbury, 2012.

Cobb, John. *Is It Too Late? A Theology of Ecology*. Beverly Hills, CA: Benzinger, Bruce & Glencoe, 1972.

Coetzee, J. M. "Notes on Issues Raised by Matthew Calarco." In *The Death of the Animal: A Dialogue*, edited by Paola Cavalieri, 89–92. New York: Columbia University Press, 2009.

Davis, Ellen. *Scripture, Culture, and Agriculture: An Agrarian Reading of the Bible*. Cambridge: Cambridge University Press, 2009.

Deane-Drummond, Celia, Rebecca Artinian-Kaiser, and David L. Clough, eds. *Animals as Religious Subjects: Transdisciplinary Perspectives*. New York: Bloomsbury, 2013.

Deane-Drummond, Celia, Sigurd Bergman, and Markus Vogt, eds. *Religion in the Anthropocene*. Eugene, OR: Cascade, 2017.

Derr, Thomas Sieger. *Ecology and Human Liberation: A Theological Critique of the Use and Abuse of our Birthright*. Geneva, CH: World Student Christian Federation Books, 1973.

Derrida, Jacques. "The Animal That Therefore I Am." *Critical Inquiry* 28.2 (2002) 369–418.

Gross, Aaron S. *The Question of the Animal and Religion: Theoretical Stakes, Practical Implications*. New York: Columbia University Press, 2015.

Habel, Norman. *Readings from the Perspective of Earth*. London: T&T Clark, 2000.

Habel, Norman, and Peter Trudinger, *Exploring Ecological Hermeneutics*. Atlanta: SBL, 2009.

Haraway, Donna J. *The Companion Species Manifesto*. Chicago: Prickly Paradigm, 2003.

———. *When Species Meet*. Minneapolis: University of Minnesota Press, 2008.

Hobgood-Oster, Laura. *Holy Dogs and Asses: Finding Animals in Christian History*. Urbana, IL: University of Illinois Press, 2001.

Johnson-DeBaufre, Melanie, Catherine Keller, and Elias Ortega-Aponte, eds. *Common Goods: Economy, Ecology, and Political Theology*. Fordham, NY: Fordham University Press, 2015.

Linzey, Andrew. *Animal Gospel: The Christian Defense of Animals*. London: Hodder & Stoughton Religious, 1998.

———. *Animal Rights: A Christian Perspective*. London: SCM, 1976.

———. *Animal Theology*. London: SCM, 1994.

———. *Christianity and the Rights of Animals*. New York: Crossroad, 1987.

———. *Creatures of the Same God: Explorations in Animal Theology*. New York: Lantern, 2009.

———. *Why Animal Suffering Matters: Philosophy, Theology, and Practical Ethics*. Oxford: Oxford University Press, 2009.

MacKinnon, Mary Heather, and Moni McIntyre, eds. *Readings in Ecology and Feminist Theology*. Kansas City, MO: Sheed & Ward, 1995.

McDaniel, Jay B. *Of God and Pelicans: A Theology of Reverence for Life*. Louisville: Westminster/John Knox Press, 1989.

———. *With Roots and Wings: Christianity in an Age of Ecology and Dialogue*. Maryknoll, NY: Orbis, 1995.

McFague, Sallie. *Models of God: Theology for an Ecological, Nuclear Age*. Philadelphia: Fortress, 1988.

Merleau-Ponty, Maurice. *Phenomenology of Perception*. London: Taylor & Francis, 2008.

Merleau-Ponty, Maurice, Claude Lefort, and Alphonso Lingis. *The Visible and the Invisible: Followed by Working Notes*. Evanston, IL: Northwestern University Press, 2000.

Moltmann, Jürgen. *God in Creation: A New Theology of Creation and the Spirit of God* . San Francisco: Harper & Row, 1985.

Moore, Stephen D., ed. *Divinanimality: Animal Theory, Creaturely Theology*. New York: Fordham University Press, 2014.

Northcott, Michael. *Place, Ecology and the Sacred: The Moral Geography of Sustainable Communities*. New York: Bloomsbury, 2015.

———. *A Political Theology of Climate Change*. Grand Rapids: Eerdmans, 2013.

Pinches, Charles Robert, and Jay B. McDaniel, eds. *Good News for Animals?: Christian Approaches to Animal Well-Being*. Maryknoll, NY: Orbis, 1993.

Ruether, Rosemary Radford. *New Woman, New Earth: Sexist Ideologies and Human Liberation*. New York: Seabury, 1975.

Rad, Gerhard von. "The Theological Problem of the Old Testament Doctrine of Creation." In *The Problem of the Hexateuch and Other Essays*, 131–143. New York: McGraw-Hill, 1966.

———. *Wisdom in Israel*. Nashville: Abingdon, 1972.

Santmire, H. Paul. *Brother Earth: Nature, God and Ecology in Time of Crisis*. New York: Thomas Nelson, 1970.

Schaefer, Donovan O. *Religious Affects: Animality, Evolution, and Power*. Durham, NC: Duke University Press, 2015.

Scott, Peter. *A Political Theology of Nature*. Cambridge: Cambridge University Press, 2003.

Urbanik, Julie. *Placing Animals: An Introduction to the Geography of Human-Animal Relations*. New York: Rowman & Littlefield, 2012.

Waldau, Paul, and Kimberley C. Patton, eds. *A Communion of Subjects: Animals in Religion, Science, and Ethics*. New York: Columbia University Press, 2009.

Webb, tephen H. *On God and Dogs: A Christian Theology of Compassion for Animals*. New York: Oxford University Press, 1997.

White, Lynn. "The Historical Roots of Our Ecologic Crisis." *Science*, 155:3767 (1967) 1203–7.

FAMILY AND HOME

The Cruciform Lap

Suffering and the Common Life of Cats, Chickens, and Couples.

—TREVOR GEORGE HUNSBERGER BECHTEL—

I should be writing this essay, but the cat is on my lap. During the ten years that Tiamat the cat lived with my wife Susan and I a rule formed from common practice and I am now respecting that rule by typing these letters one at a time with the only finger that can reach my keyboard. The rule is that if the cat is on your lap, you should not do anything to disturb the cat. If you need more coffee, or tea, or a pen, or if the laundry needs to be put in the dryer and the cat is on your lap, then the other human being has to get it for you. Herbert McCabe suggests that rules are what rational people would do anyway[1] and in this case the suggestion holds. If you lived with a cat who could be affectionate, but was not affectionate very often, then you would maximize time with the cat on your lap too. Now that Susan and I live with Neko, the accidentally eponymous cat named after Neko Case[2] and who is demanding of our affection and laps at every turn, productivity in our household has plummeted. This is the story about how I have tried to understand, and come to terms with, the most decisive time I held Tiamat on my lap.

When I entered graduate school, aware of the absences mental and physical that I was about to subject Susan to, I decided to get her a cat for her birthday. Up until this point in my life I had performed a strong dislike

1. McCabe, *Law, Love and Language*, 53.

2. Neko is named after the singer Neko Case. We only learned later that *neko* is also the Japanese word for cat.

of cats very publicly. The roots of my dislike were found in the symptomatic reaction my eyes and nose endured every time I visited a space were a cat lived, but I'd heard you could overcome a reaction to one animal through constant exposure. I had a friend take Susan out for supper and arranged for them to come back to her friend's house afterwards. I asked several of Susan's other friends to gather at the friend's house. They all brought gifts of cat accessories and I had a certificate good for one cat from the Tree House, a shelter on Chicago's north side. We did this five months after Susan's birthday. It was the most successful surprise party I have managed. A space of pure gift and abundance was created in our mutual life, and now we just needed to find the cat who would inhabit it.

We visited the shelter and tried to find a cat we both liked. Susan warmed to a fat cat but when I visited that cat he jumped away and carried by the force of his weight crashed clumsily to the floor. I warmed to a small white three-legged cat with health conditions, holding the idea that if I didn't adopt this cat no one probably would. When Susan visited this cat he also lost his balance in what seemed a drugged stupor. However, this combination of interest had an unintended effect on the shelter staff and they brought us to see Mootsie, a cat convalescing from the surgeries needed when she entered the shelter. Mootsie was brought to the shelter with a recessed eye and a badly injured leg acquired in some kind of epic battle. She was spayed quickly, and a flap of skin hung from her belly. The leg should have been amputated, it was cut to the bone, but the surgeon reconsidered and opted for patience, which was rewarded. When we meet her a week before she was able to be adopted, she moved right to Susan and sat on her lap. This convinced both of us, and the shelter, to forego the ordinary rules and invite her into our home.

This easy deliberate moment of lap sitting was a total bait and switch. For the first week that she lived with us we hardly saw her. Food would disappear from her bowl and her litter box was used, but for days we couldn't find her. We looked. We lived in a tiny apartment—just over 200 square feet—but we couldn't find her. Finally, I noticed that some of the books on my bookshelf were out a little bit over the edge. Knowing that I would never do that I investigated and that she had created a small hiding place behind some green books. Unobservable from any angle, she made clear by her hiddenness that interaction with us would happen on her terms.

It was obvious to us that Mootsie was a bad name for this deliberative and persuasive cat. Susan wanted to name her for a goddess and we settled on Tiamat. I appreciated the resonances to the Enuma Elish. Tiamat, the cat and the goddess, both had bellies obviously split open, and both went on to create a new world. We lived with Tiamat for fifteen years. She was a cat with a huge personality, vocabulary, and impact. She was not friendly, but she could be very loving. She despised strangers unless she could intuit that they did not like cats in which case she would demand their attention and affection. A friend who stayed with us for a few months lived in fear of the staircase in our house; if he ascended or descended at the same time as Tiamat he would be rewarded with a swipe. One night when we were away he was lying on the couch watching TV, and Tiamat chose to sit on his chest. He didn't move for hours, bound by fear, surprise, and love. The same thing happened to our niece Evelyn, who stayed with us for a few weeks before beginning high school. After a lifetime of wanting to interact with Tiamat and being rejected, Tiamat chose to sit on her lap.

I could tell many stories about Tiamat. The time my parents foolishly gave her catnip resulting in a set of attacks against their collie—easily ten times bigger than Tiamat—so vicious that they needed to lock her in a spare bedroom overnight. The time we took her to a cabin and she chased the squirrels that lived in the walls all night long. When I tried to let Susan get some sleep by taking Tiamat for a walk she ran ten feet up the side of tree, leaving both us confused about the gravity of leashes. The time at Susan's parents when, ready to leave, we couldn't find her anywhere. She was hiding in the rafters in a storage room; don't ask me how she got there. To get her out I needed to yell instructions at my father-in-law; something I still can't imagine doing under any circumstances. But the story that I like the most is the one with Susan or I doing something that unexpectedly offended her and listening to the lecture that ensued. She was free with hisses and swats, but she would often also seek to explain herself in a series of chirps, mews, meows, and barks. She had a huge vocabulary. Once she decided to lecture a bird on a wire in our backyard in Chicago. I was on the upstairs balcony watching. After a minute or two of enduring a steady stream of invective, the bird started to respond. Soon another bird joined the first on the wire. Then another. Without regard to species, bird after bird lighted on the wire and took up the chorus against Tiamat. This continued until there were no fewer than thirty birds. Tiamat, slowly, started to back up and then turned and slinked up the stairs to the safety of our apartment. It was the only time I ever saw anyone get the better of her.

Tiamat was happy for food and protested vehemently if there was not always food in her bowl, but food did not make her happy. She ate only dry food. Reflecting on this now, I realize that her preference for dry cat food obscured the fact that she, like all cats, was an obligate carnivore. Susan and I do not call ourselves vegetarians but we very rarely have meat in our home except, I remind myself, for the cat's food. The few times Tiamat chose to eat something other than dry cat food it was always grass, and although she used it almost exclusively as an emetic, it still seemed like her food of choice was grass and the small brown pellets, not obviously meat in any way, were a kind of ascetic practice as much as a succulent meal. Once, Tiamat became intolerant of her current brand of cat food and we got some chicken prepared by friends for cat consumption. Their cats ate only fresh chicken, prepared, frozen, and then thawed. Tiamat had no interest. When we needed to give her medicine we couldn't put it in a treat or mix it in with wet food because she didn't eat either. We were happy to have her around when we were eating because she had no interest in our food. Strangely, Tiamat did prefer to drink out of a water glass that Susan had first drunk from, and was constantly claiming Susan's water as her own. Once, at a vet's suggestion, we put her on a diet. She had put on a bit more weight than she needed. She became agitated, not just with us, but also with herself, and engaged in compulsive grooming. This was very stressful for all of us and relatively quickly we gave in. There was no give and take with her, but only give and, perhaps, receive.

The most decisive story about how we received love from her came during a difficult time in our lives. In our late twenties Susan and I were trying to have children and it wasn't going well. We sought progressively more and more invasive medical help from both alternative and conventional sources. One night we were lying in bed and Susan broke down crying from the stress of it all. Tiamat came, more or less from out of nowhere, and unpredictably moved up to Susan's face. To our great amazement she started to lick the tears from off of Susan's cheek. Once Susan was clean, Tiamat nuzzled her head into Susan's neck and remained, comforting us both beyond measure.

In her later years, Tiamat's health declined. Urinary tract infections became frequent. Her good eye was visited by a variety of ailments, testing our patience, ability, and capacity to provide care. Treatment was expensive and required us to travel two hours to Columbus on occasions, a trip that none of the three of us enjoyed. She required drops and regular testing that we administered at home, risking our own bodily integrity to guarantee hers.

We moved from Bluffton, Ohio to Ann Arbor, Michigan, anxious for how she would fare in new space. Somehow the eye problems cleared up, and, more surprisingly, Tiamat recalculated the relative benefit and cost of spending time on laps. My father, now eighty-five, has left behind a lifetime of thrifty thermostat settings because his aged body is now always cold. I think a similar thing happened to Tiamat, and she had the advantage of being able to warm up by sitting on laps. Cat sitters who knew her in these last couple of years would regularly report, to our amazement and incredulity, that she spent much time sitting on their laps.

So, our relationship with Tiamat was not one of exchange, of her tolerating or being affectionate towards us in exchange for food, or shelter, or safety. This was mostly at her initiative. She offered affection, attention, or engagement entirely on her own schedule. As we observed that this was how it was going to be we also adjusted. She became more of her own person, a third member of our family, and less of a pet. This meant that Susan and I related to her as a distinct person, but also that our relationship was mediated by having her in our lives and in our home. This was work that Tiamat did by shaping her interactions with us in a particular way. Had Tiamat evinced the easy loyalty and affection of a dog (not there's anything wrong with dogs), it would have been harder for Susan and I to learn from her.

And so, while I want to credit Tiamat with shaping most of the dynamic of exchange in our relationship to her, I need to also credit Susan and myself. For it is true that the kind of openness that we had in seeking her out created the space that she could then fill and expand. And, were we not willing to welcome her personality as authentic and worthy of respect, we would not have been able to learn.

The Cat as Priest

This dynamic is not a particularly easy set of ideas to put into words. I turn now to considering a set of connected reflections by Rowan Williams and Eugene Rogers to give a theological account of the relationship between Susan and I but more importantly how Tiamat shaped that relationship.

> To be formed in our humanity by the loving delight of another is an experience whose contours we can identify most clearly and hopefully if we have also learned or are learning about being the object of the causeless loving delight of God, being the object of God's love for God through incorporation into the community of God's Spirit and the taking-on of the identify of God's child.[3]

Williams here is discussing the role of the celibate person in witnessing to the love of married people by analogy to the perichoretic love of the Trinity, but these words are better than any I can find to discuss the role of Tiamat in my life. Tiamat was a witness to the love that Susan and I share in the sacrament of marriage. I believe that the depth and breadth of my love

3. Williams, "The Body's Grace," 113.

for Susan and her love for me was expanded and confirmed by Tiamat. As I would relate to Tiamat, Susan would see someone interested in engaging in the full spectrum of seeking to be in relationship with a cat like Tiamat. Because Tiamat's engagement was so fickle, this commitment seemed, and was, more obvious and authentic. We think that we can know someone by how they treat animals and by how animals treat them. This is true in a general, and in a first meeting sense. But it is also true, and true in a broader and deeper way over time.

Eugene Rogers builds on the account offered by Rowan Williams and adds insights from Paul Evokimov in his book *Sexuality and the Christian Body* in order to draw out an account of God's love, desire, and the indirect and direct ways that marriage and monasticism attain God's love. Rogers suggests that the monk, or priest, or in Sebastian Moore's term the truly celibate person, participates directly in God's love. Married spouses learn about God's love through the love of each other. Each kind of love, each kind of participation is important and as Rogers avers below, needed. There is also for Rogers (and Williams) an analogy of the monk and the married to the Trinity.

The Father and the Son love each other in a mutual, self-emptying love that has as both its nadir and zenith the crucified Christ hanging on the cross. This constant emptying of love towards the other is facilitated by the work of the Holy Spirit, who brings these gifts of love back and forth and forth and back from the one to the other. Trinitarian loving relationality is dependent on three persons in this model. The analogy to the monk and the married is both direct and indirect. The monk participates directly, emptying himself of sexual relation, in order to witness to the love of the married person. The monk's work directly parallels the work of the Holy Spirit. The married participate indirectly, loving each other and in that love seeking to discover, with the help of the monk, God's love. Here is how Rogers works this out. He begins by quoting Evokimov,

> "What the monks attain *directly*, the spouses work out *indirectly*, and their means is the sacramental sphere of grace. The one through the other they look at Christ." It is this transformation wrought by others in the community of God and the Church that establishes the true, sacramental, sanctifying orientation for Christians' physical desires. Both these forms of community—monasticism and marriage—require time to complete the transformation of human beings by the perceptions of an other. Both the married and the monastic need somebody who loves them to call them on their faults from whom they cannot easily escape. The transformation is not only, or even primarily, the

> experience of falling in love (*eros*), but that is the intensity and the clue to the importance of something else: the experience of living with someone, the neighbor, who won't leave one alone (*agape*).[4]

This transformation, this not being easily able to escape, this neighbor who won't leave me alone, all of these things resonate with how I experienced Tiamat. Even, I suppose, in her life, lived apart from nature with an ascetic diet and cloistered in our house, Tiamat was always a good monastic priest. I have sought to describe so far many of the different ways in which my relationship with Tiamat, particularly as I have understood it vis-a-vis my relationship to Susan, has shown me God's love. The final proof I have of this is found in the last moments of Tiamat's life.

It had been a bad weekend in December. Tiamat lost control of her bowels, could hardly walk, and was visibly exhausted. Susan and I both took turns holding her on our lap, saying goodbye. On Monday, Kathy the home visit vet shows up and asks if one of us wants to hold her. I say yes, that I'd like to be with her, to be a witness to her shaking off this mortal coil. I take her into my lap and immediately begin bawling uncontrollably. Kathy the vet is nearby, waiting for me to stop shaking so she can administer a lethal dose of some drug that will end Tiamat's life. I can tell that Kathy is getting ready to take over, that the amount of emotion I'm physically expending is starting to become not okay. But I get the shaking, if not the bawling, under control. The first needle goes in, comes out. The second needle. Too soon I realize that the lump of flesh I'm holding in my hands is no longer Tiamat in any meaningful way. I realize that my lap has disappeared entirely and so I get up. Kathy leaves. We buried Tiamat in the fortunately thawed soil and planted wildflowers around her grave the next spring. The wildflowers are not in bloom yet as I write this, but I still stop routinely at the stacked flagstone grave marker and remember my friend, my priest. It's still strange to me that I chose the date of Tiamat's death, that it would have been cruel not to end her suffering. I'm still angry that the university I worked for then, which has a practice of sending out all-campus emails at the death of an employee's family member, refused to send an email out about Tiamat.

Two nights after she died I have one of the strongest dreams I can remember. In the dream, I am holding Tiamat in my lap and petting her and I awake with the touch of her fur on my hands. It is the only time I have felt something in a dream, and although it is faded, the memory of that sensation persists even now. It was, and is, as if Tiamat returned to me to grace my hands once more with the feeling of her fur, the weight of her on

4. Rogers, *Sexuality and the Christian Body*, 83.

my lap, the lap that had disappeared in the moments her life ended. And, my lap wasn't the only thing that disappeared that gray day in December. Susan and I came to realize over the next days, weeks, and months that our marriage had a new challenge. What did it mean for us to be married, to share the same space, now that we would not have Tiamat to mediate that space or that relationship to us? It required a whole new pattern of life. That space is occupied, very differently, by Neko, but there are still gaps that will never again be covered, at least in this life. And so I understand the indirectness of the analogy to the Trinity in Williams and Rogers quite well, and I wonder how well Tiamat experienced the directness of it. Rogers says,

> The Trinity is the paradigm or perfect case of otherness as an exchange of gift and gratitude, that from which all other cases of gift take their (analogically but not historically or empirically) derivative meaning. The technical name for "gift" in Christian theology is "grace," and gift in that sense is especially appropriate to the Holy Spirit. What is grace? Is it pretty sunsets, serendipity, any kind of positive regard? It is the particular characteristics of God as Father, Son, and Spirit that define grace as a particular characteristic of this God. In Williams's phrase, it is first of all that "unconditional response to God's giving that God makes in the life of the Trinity," and secondarily, through God's turning to God's human creatures, it is God's incorporation also of them into the fellowship of gift and gratitude.[5]

Two things become clear here. The more trivial is that I may be reading Rogers, and perhaps Williams, somewhat against themselves as Rogers is clear that it is "human creatures" that are incorporated into Trinitarian perichoresis. However, I take this to be trivial because Rogers himself reads Barth and Aquinas somewhat against themselves in *Sexuality and the Christian Body* but sees this reading as the work of theology, to engage in a kind of thick description that creates its own argument. The other is that "otherness" need not be a barrier to experience the gift and gratitude of love. Grace is an unconditional response. Like many humans, I learned first about unconditional love from my mother. But the unconditional response of a cat is a different thing and probably better teaches, especially to the comfortable, the surprise and joy, the concern and weight, of love.

I have sought to describe many of the different ways my relationship with Tiamat has shown me God's love. Following Rogers, I've suggested that Tiamat was a monastic priest to me and that she witnessed to the analogy to God's love in my marriage, and showed me the presence of God in her

5. Ibid.,198.

own life. In the end, I take this last clause to be self-evident as it seems obvious that many animals seek after their own purpose naturally when not obstructed from doing so. I have always taken the "waiting" in Romans 8 as the "waiting" of someone who already gets it and is waiting for the others to catch up. What I haven't done, and what is more difficult, is to articulate the role that Susan and I played in Tiamat's life. Were we necessary to her? Were we able to call her on her faults? To answer these questions will require more than a few shifts of perspective.

The Story of Domestication

One of my first teaching assignments was a preexisting course on "The Ethics of Intimacy" at the Episcopal Seminary in Chicago. The course syllabus included both Rogers and Stephen Webb's *On God and Dogs* and conversations from that course have had a great deal of impact on this essay. In particular, one student noticed that in Rogers's model there is a positive role for the truly celibate person, and for married people both gay and straight, but no redemptive account of singleness that is not chosen. We had drawn a model of the relationship Rogers describes on the chalkboard, including his modeling of the Trinity and sacramental participation in God's life, and she suggested that the only location for the single person was suffering at the cross. When the Trinity is an analogy to marriage the single person at the cross is emptying themself, but only in witness to their own pain. The single person could hope for redemption at the eschaton, but for now, in Paul's words from Romans 8:19, she "waits with eager longing for the revealing of the children of God . . . [but is] subjected to futility."

The analogy between the Trinity and monks and married people is compelling to me because it makes sense of the positive role that Tiamat played in my life. It is easy for me to see cats as analogous to the Holy Spirit. Cats are domesticated creatures but we often joke that it is they who have domesticated us. In any event, it is easy to see the spaces in which cats, especially cats like Tiamat, resist complete domestication. It is easy to understand the cat alongside the Holy Spirit in their common resistance to domestication. But what if the role that I played in Tiamat's life was only negative? What if by forcing her to live with me I was denying her true feline flourishing? What if, like my perceptive student, she felt less like the Holy Spirit and more like someone abandoned on the cross?

The question here is one of the theological meaning of domestication, particularly the redemption of suffering for domesticated creatures. Is life for an indoor cat luxury or captivity? What is the meaning of food for a domesticated cat? Stories of well-fed cats who terrorize the small fauna of suburbia are legion. Are they acting out an essential nature that I have denied to my cats? Is a day spent sleeping in the sun a joy, a basking in God's grace, or the symptom of depression? Is the cat, calm, cool, and in total sovereignty over all, who suddenly leaps into action—attentive and engaged—driven to decisively chase cherubim wherever they appear? Or is that cat insane, driven to madness by boredom? What is the self-emptying behavior that a cat engages in as an unconditional response to God's love? The easy answer to these questions is that we can't know, that if cats have a theology it is surely apophatic. But a more articulate answer begins in the consideration of the life of another domesticated creature.

A few Fridays ago, earlier than I'd care to admit, I went over to my friend Dave's to murder some roosters. Five roosters met their end in the crispness of that morning, but they died as well as deliberate killing could end a life. In fact, their deaths convinced me that a good death is a possibility for farm animals, a question I had wondered about since reading Jonathan Safran Foer's *Eating Animals*. It may be that a truly good death is only possible to animals killed a small number at a time given the different ways that the knife needs to be wielded to achieve economies of scale—but a good death is possible. Meat is murder, but humans and the animals that we live with share a peculiar and highly asymmetrical relationship.

We have historically thought of domestication as a process of humans controlling and forcibly domesticating other animals. Recent science suggests that the process was probably much more mutual, and not just for cats. Domestication is the process of learning to live with another species in a *mutually* beneficial arrangement. So, at some point several chickens, but also some sheep, horses, dogs, and cats left their lives in the wild and joined the new experiment of life with humans. This has been a remarkably successful strategy. There are many more domestic animals than wild. And while many domestic animals are neotenous they also have many signs of increased social intelligence. Domestic chickens, for instance, make many more sounds than their wild counterparts. In general, living in a bigger group demands more brain power, and therefore more speech than living in smaller groups. Cats who live with humans also tend to increase their vocabulary when talked to, even though humans have yet been able to discern a role for most cat vocalizations in cat-to-cat communication. Furthermore, biblically, domestication is a part of God's creative work; domestic creatures show up in both creation accounts.

All summer long those roosters abundantly frolicked, hung out, ate, drank, roosted, pecked, clucked, and generally lived a good life. This life was only possible for these individuals if it ended with their becoming meat. They would not have remained in existence otherwise. My friend Dave buys roosters because they are cheaper than hens, and then cares for these animals, feeding them the right amount of food and supplements. He gives them limestone for their gizzard (without it they couldn't "chew" their food). And he knows how to kill them well. The rooster is placed head first into a killing cone, basically a traffic cone turned upside down and nailed to a fence post. The head comes out the bottom of the cone and gravity compresses the roosters body inside the cone. This compression is comforting to a rooster. He relaxes, perhaps not unlike a cat on a lap. A decisive cut to the jugular vein and the rooster hopefully passes quickly into shock from quick blood loss. The blood drains and with it the creature's life, again hopefully before the shock passes. Lots can go wrong here, but it doesn't need to. I didn't take the knife that morning partly because it wouldn't respect the rooster; their death came about more easily given Dave's experience. Even so, Dave's cut was a bit less decisive for the first of five roosters, and, as the rooster was dying he craned his neck up and met my gaze.

It is quite the thing to look at someone in the eye as they are dying. I remembered in the killing of this rooster the last time I'd watched a creature die. But my relationship to my friend of fifteen years was different in intensity to my relationship with the rooster I had only met a couple of times, and honestly couldn't tell apart from his compatriots. The rooster was in many senses my neighbor, in his dependence on Dave and for the weekend that Dave was up north, on me, in his geographical proximity to me, and in my connectedness to him through Dave. But the rooster was not my friend. I may have been intellectually interested in such a relationship, but the rooster was not, and practically I'd betrayed the roosters more than once.

One summer long weekend Dave had asked me to care for the roosters, to feed them and to provide some limestone and water. The weekend came and I'd been thinking about the roosters all weekend long. On the holiday Monday I'd been cold all day, putting on layer after layer to try and keep warm. And then I realized it. In a second my body temperature rose at least ten degrees and I was flush with heat. "Ugh. I forgot to feed them." They'd been more than forty-eight hours without fresh food and water. I'd been entrusted with their care, and I'd failed to be a good neighbor, let alone a friend. I rushed, with Susan in tow, the short drive to the roosters. I ran up to their pen. They were okay. I fed them, watered them, and confessed my sins to Dave who was just coming home from the weekend away. I think he's forgiven me, but he hasn't asked me to care for the roosters again.

Anyway, it is quite the thing to look at someone in the eye as he is dying. I looked at my neighbor the rooster and held his gaze and thought two things. This is murder, what we are doing, but it's worth it. That night I went home and devoured the chicken leg Dave had given me, confident that he'd been given an abundant and good life and that he now was giving me a good supper. After he kills the rooster, Dave, rather than defeathering the bird, removes the entire skin. This happens, at least in his hands, quite easily. The requisite cuts and pulls and the skin comes away from the body quickly. The rooster is transformed into chicken meat. Both Tiamat and the rooster quickly moved from life to flesh. This was obvious to me, but it is interesting to reflect on the different ways that recognition worked in each case. When Tiamat died she became unrecognizable and, since I have very little experience with dead cats, I did not have a ready category for the lump of flesh in my hands. When the rooster died and was skinned and became meat my recognition actually increased. Although Susan and I very rarely have meat in our home, I have much more experience with chicken meat than I do with live chickens. I recognized the meat perhaps even more than the rooster. It is also revealing to think about the connection between Tiamat's death and the rooster's death. I do not kill the chickens that my cats eat. In fact the kibble that they eat is hardly recognizable as chicken. But the cat's life, even more than mine, requires the death of the chicken. This is, of course, a recent phenomenon. For millennia the cats who lived alongside humans mostly found their food in the rodents who had also found our pattern of settlement opportune. Food was work for these cats.[6] But now, humans like Susan and I invite cats to leave shelters and come and live with us and we provide chicken for food. We do not ordinarily expect that a story ends well with one character killing the other.[7] Still, I believe that this is exactly the story that we need to tell about how we live with domesticated animals.

The Gospel of All Creatures

This is way that this story is told by those sixteenth-century German Mystics and Anabaptists taken with another analogy that binds creatures to the divine, the strange and wonderful trope of the "Gospel of All Creatures,"[8]

6. See Bradshaw, *Cat Sense,* for a useful discussion of the human-cat relationship.

7. I'm grateful to Darrin Snyder Belousek for this formulation, which he employed in the question and answer session after a paper I gave on Toola the sea otter at the Society for Christian Ethics. See Bechtel, "Re-Imagining Personhood," 2017.

8. Rupp, "The Gospel of all Creatures," 493–94

and in particular this idea which one can find across authors who employ the trope.

> (Paul) says further that the gospel I preached to you is preached in all creatures. God created all creatures in five days so that they can be used by the human, who was created on the sixth day. Then the creature has its rest. But the human being was not created to remain a human being on the sixth day, but that he would come to the seventh day, indeed that he become godly, or divinized, and come to God. That is then the appropriate human rest or true day of celebration. And indeed the means by which all creatures come to be useful to the human is suffering. One kills, cuts and cooks and the creature holds still and suffers for the sake of faith. And just as an animal is not useful to the human for food unless the body dies, so no human becomes blessed, who does not die for Christ's sake.[9]

Leonhard Schiemer may or may not have a killing cone in mind when he remarks on the creature that "holds still and suffers for the sake of faith" but he does suggest that the use of creatures by humans is oriented towards giving the creature their rest. For domesticated creatures, in fact for all creatures, the way to rest is through suffering, by following Christ and yielding ourselves to the other and to God. The end for the creature in the Gospel of All Creatures is suffering, but this is not an empty suffering, it is instead set inside the context of God's justice and order.[10] It is the context that makes all the difference inside the awesome world of the Gospel of All Creatures and it is a context here that connects suffering and love in an eschatological or even apocalyptic assessment of how humans are saved. Pilgram Marpeck offers perhaps the most forthright expression of the idea that the witness of creatures is sufficient to ground human understanding of God's love.

> Yes, even if a dog or a cat were to proclaim the gospel as a testimony, throughout the unbelieving world and deliver it into

9. Liechty, ed., *Early Anabaptist Spirituality*, 32.

10. This trope came into Anabaptism through Thomas Müntzer, who reflects on the importance of paying attention to creatures this way. "I have not heard from a single scholar about the order of God implanted in all creatures, not the tiniest word about it; while as to understanding the Whole as a unity of all the parts those who claim to be Christians have not caught the least whiff of it—least of all the accursed priests" (Matheson, ed. and trans., *The Collected Works of Thomas Munzter*, 357). By insisting that the Whole—God's divine order—was evident in the creatures and by insisting that all of the scholars, priests and doctors of faith had entirely missed this essential of faith, Müntzer here captures ideas resonate in the German Mysticism of thinkers like Meister Eckhart and the author of the *Theologica Deutsch.*

> repentance an improvement, who could declare it wrong? For everything that leads to godliness is good, and not evil, for all visible creatures are placed in the world as apostles and teachers (Job 12). If such mute creatures could speak, Christ's sending the apostles to elucidate or preach the gospel would have been unnecessary.[11]

Marpeck's modern interpreter Stephen Boyd argues that Marpeck intends to embed suffering into his treatment of the "Gospel of all Creatures" in this quote by reference to Job 12.[12] In Job 12, Job indicts Zophar with being unable to understand that which even simplest creatures can understand—the hand of God in Job's suffering. Marpeck continues the line of thought evident in Schiemer by suggesting that the trajectory from life to suffering to rest acts as a witness and testimony to God's larger purposes.

The Cruciform Lap

To this point I have reflected on the life and death of Tiamat, a cat I was very close to, who had a priestly role in my life and marriage, and who I understand to have been an analogous presence to the Holy Spirit; and the shorter life and death of a rooster, who became my food. What is the analogy that connects these concrete lives to the life of the Trinity? The connection of food and sacrifice and the mediating role of the priest in eucharistic theology provides one possible direction. Another possibility is in the resurrection, which binds at least Christ and possibly Tiamat together. After all, in my dream, there were no barriers to my tactile encounter. Perhaps Tiamat visited me after her own three-day sojourn. And, the self-emptying moment in the lives of Christ, my cat, and the chicken provides another direction, but as I hinted to above, I am anxious about ascribing self-emptying behavior to cats or chickens. This is not because those creatures do not have selves; I am convinced that they do. I am anxious instead about thinking that I can know that their self is choosing the emptying that gives their lives the meaning I want it to have. My anxiety is, of course, particularly strong given the legion of straight white male theologians eager to find self-emptying behavior in oppressed people, be they black lives who mattered before being killed, women whose worth should have prevented their rape, or any of a plethora of other examples. This anxiety is all about context and, I believe,

11. Klassen and Klaassen, eds. and trans., *The Writings of Pilgram Marpeck*, 56.
12. Boyd, *Pilgram Marpeck*, 77.

meaningful choice. This is the concrete and spiritual "lap" that I have been holding throughout this essay.

When Tiamat chose to sit on Susan's lap, to eat the food we provided, or to lick away her tears, was she also choosing to become a priest? Was she choosing to witness to God's love? I believe that she was, even though this is in some ways a relatively new role for domestic cats generally, to eat chicken rather than mice, and to be pet companions of humans rather than pest controllers for them. Although I could not ask her what she wanted and hear a response in language—the cat being no different than Wittgenstein's lion—cats are notoriously good at making their intentions and desires known even if their humans cannot understand them.

The question of the chicken's choice may be more difficult, but in a yard like Dave's it may also be easier. Given the opportunity to peck and roost and bathe and socialize the life of a backyard chicken is mostly calm and even happy. Bertrand Russell may be right when he says, "The man who has fed the chicken every day throughout its life at last wrings its neck instead, showing that more refined views as to the uniformity of nature would have been useful to the chicken,"[13] but the chicken may end up caring more for the divine order of the Gospel of All Creatures than Russell's uniformity of nature. The chicken may be disappointed that humans have shifted their cat's diets from mice to chicken, but if the original "choice" to live with humans holds then the chicken may not be disappointed with the increase in chicken dying and its consequent increase in chicken living and therefore flourishing.[14] After all, "Chickens also have a sense of the future. Given a choice between receiving a small amount of food immediately or a larger amount in the future, they will choose the latter, demonstrating self-control and the capacity to delay gratification."[15]

Most importantly, to take these connections with the seriousness I am seeking to ascribe to them means that domestication has in some of its mutual beneficence not just a common flourishing but also some symmetry. And so the space on my lap, I hope, is an open one, seeking to hold not just the love I am given, the sacrifice I require and which saves me, and the suffering all around me, but also the possibility that I am like the chicken

13. Russell, *The Problems of Philosophy*, 35.

14. In this essay I am considering only the connections in the concrete lives of a particular chicken, a particular cat, and a particular couple, and can therefore sidestep for now the problem that the chicken I knew was privileged beyond measure in comparison to his peers on a factory farm.

15. Hatkoff, *The Inner World of Farm Animals*, 22.

and the cat, dependent on them for who I am and for my own capacity to participate in the rest of God's divine order.

Bibliography

Bechtel, Trevor. "Re-Imagining Personhood." *EcoTheo Review*, May 16, 2014. Accessed April 27, 2017. http://www.ecotheo.org/2014/05/re-imagining-personhood/.

Boyd, Stephen B. *Pilgram Marpeck: His Life and Social Theology*. Durham, NC: Duke University Press, 1992.

Bradshaw, John. *Cat Sense: How the New Feline Science Can Make You a Better Friend to Your Pet*. New York: Basic, 2014.

Foer, Jonathan Safran. *Eating Animals*. New York: Little Brown, 2009.

Hatkoff, Amy. *The Inner World of Farm Animals: Their Social, Emotional, and Intellectual Capacities*. New York: Stewart, Tabori & Chang, 2009.

Klassen, William, and Walter Klaassen, eds. and trans. *The Writings of Pilgrim Marpeck*. Kitchener, ON: Herald, 1978.

Liechty, Daniel, ed. *Early Anabaptist Spirituality: Selected Writings*. Mahwah, NJ: Paulist, 1994.

Matheson, Peter, ed. and trans. *The Collected Works of Thomas Munzter*. Edinburgh: T&T Clark, 1988.

McCabe, Herbert. *Law, Love and Language*. London: Continuum, 2012.

Rogers, Eugene F. *Sexuality and the Christian Body: Their Way into the Triune God*. Oxford: Blackwell, 2000.

Rupp, E. Gordon. "The Gospel of all Creatures." *The Bulletin of the John Rylands Library* 43 (1961) 493–94.

Russell, Bertrand. *The Problems of Philosophy*. Oxford: Oxford University Press, 1912.

Webb, Stephen. *On God and Dogs*. Oxford: Oxford University Press, 1998.

Williams, Rowan. "The Body's Grace." In *Moral issues and Christian responses*, edited by Patricia Beattie Jung and L. Shannon Jung, 106–15. Minneapolis: Fortress, 2013.

The Story of Max

—John Berkman—

Should We Get a Dog?

Max, our Retriever-Great Dane mix, would come into our family's life in the summer of 2016. After finding him through Petfinder, we would drive the five hours to meet him, and on that first meeting, formally adopt Max. Our life with this rehabilitated rescue dog would allow us to enter the life of a heroic dog rescuer and would teach our family lessons about dogs and our family that we could learn no other way. But most of all, it was a love story of sorts, with the messiness such stories entail.

That's the basic story. But the significance of Max in our family's life, and the eventual heartbreak to come, can only be understood in light of the long and tortuous process that led to Max entering our family. It would be particularly devastating for Jack, our ten-year-old son. For five years Jack had been asking for a dog. My wife Jennifer and I knew Jack was serious about it. He would happily and responsibly care for dogs we periodically house-sat, and would sit literally for hours on hot summer days with a neighborhood dog who was lethargic from the heat, just to be in her presence.

For about three years we put off getting a dog for a variety of legitimate, or at least defensible, reasons. Then for about two years we were wait-listed with a charitable organization that trained Labrador Retriever therapy dogs (e.g., for the blind, hard of hearing, children with autism, etc.). They periodically had available what they called "career-change" dogs. A career-change dog was one that flunked out of the training program, either for behavioral or medical reasons. Despite our periodic calls to the charity, inquiring where we stood on the list, it was going on two years that they had not come through for us. And our son continued to pine for a dog.

As part of our delaying tactics, we got Jack all kinds of books about dogs, including one with details about various dog breeds. We learned about the characteristic personalities of different breeds, their typical grooming requirements, their levels of energy and degrees of playfulness, and which were easier or more difficult to train. I happily went along with everyone in the family trying to pick out the "right" breed for us.

The whole idea of researching dog breeds and selecting one was new to me. Having had a number of "rescue cats" up until then, I always figured if I ever got a dog, I would adopt a "mutt." After all, were not purebred dogs something invented by aristocrats of old to further distinguish them from the common, lowly people? Were not these aristocrats getting fancy dogs as a way of implying that not only were they superior to the plebeians, but that their dogs were as well? Did I want to participate in these historically class-based distinctions? Only in the context of "purebreds" did you get "mutts," which even today are typically considered lesser or inferior dogs. Furthermore, had not the desire for purebred dogs historically led to the health problems associated with inbreeding, which was bad for those specific dogs and dogs generally? My incipient views about dog breeds and breeding were neither well thought-out nor nuanced. They were also a misleading and reductionistic view on the history of breeding dogs. Nevertheless, these assumptions on my part, plus the appealing idea of rescuing an abandoned and needy dog, had in the past kept me from considering a "breed" of dog.

So how could I justify to myself our family getting on a waiting list for what was a purebred Labrador Retriever? I rationalized that here we would be getting the best of both worlds. We would be getting a "rescue" dog of sorts, and also getting one whose training, personality, and behavior would be very likely to make the dog relatively easy to integrate into our family, as none of us had any experience of training or living with a canine companion.

There was one other issue about getting a dog. As a long-time academic and animal advocate, it was the larger philosophical and theological question of whether having a pet was wrong in itself. Were not all animals better off living in the wild, rather than a confined life among humans, whether it be in a home or a zoo? Was not having a pet simply another way of instrumentalizing animals for human pleasure? Were they not also God's creatures, whose chief end was glorifying God through flourishing according to their kind? And did that conviction not rule out our "enslaving" other animals by keeping them as pets?

While I understood the logic of the anti-pet arguments, I always found these claims preposterous on a number of levels. First of all, the intimate relationships humans can and do have with their pets—especially I think with their dogs—is perhaps the greatest "raiser of consciousness" to the wonder and glory of God's other creatures. Second, one of the most powerful arguments of animal advocates to eliminate the cruel treatment of factory-farmed animals is to argue that if you would not do this to your dog, how can you support such treatment of other intelligent and loving creatures? Even zoos—and I am thinking of those zoos that make adequate effort to give their animals a habitat as appropriate as possible to their natural inclinations—can do a great service in introducing people to the wonders of God's diverse creation.[1] Zoos are sometimes the last hope for preservation of certain species, which can in future be reintroduced into the wild. And they educate the public on the importance of habitat preservation, making a real impact on the public's understanding of the importance of actually keeping or recreating wild spaces. Even if the "quality" of life of some of these species is reduced, they are witnesses to their conspecifics, and such sacrifice would be a real service to their kind.

I also believe anti-pet advocates are mistaken in their theological view that all species are "meant to be" in the "wild," or are better off not being in close relationships with other species, especially human beings. In fact many species live in a mutualistic relationship with another species—that is, the two species work together in some sense and both species benefit from the interaction.[2] Think of oxpecks and zebras, or plovers and crocodiles. And some species coevolve with other species, that is, each species evolving in part through reciprocal selective effects. For example, Hare and Shipman, among others, argue that thirty to forty thousand years ago wolves approached humans to initiate a relationship, and it flourished because it was indeed mutualistic.[3] Furthermore, Belyaev's silver fox experiments show that a wild species (one that is typically wary, fearful, and/or aggressive around human beings) can show significant signs of domestication

1. By "zoos" I am referring only to the great zoos—typically publicly funded and supported. With rare exception, the seasonal, for-profit zoo operations should either commit to the standards of the best zoos, or be closed. Typically, they simply do not have the will or the resources to provide appropriate habitats and food for their animals, nor do they educate the public about these species. Failing to do so, they simply degrade both the animals in their care, and the people who go to be entertained by them.

2. Much contemporary work in evolutionary anthropology focuses on the coevolution of humans with a variety of other species. For two contemporary philosophical accounts of humans living in multi-species "families," see Gaita, *The Philosopher's Dog*; Haraway, *When Species Meet*.

3. Hare, "Survival of the Friendliest"; Shipman, "Do the Eyes have it?"

in just three generations, and can be more or less domesticated in eight generations.[4]

What is even more interesting are recent anthropological claims not only that dogs became domesticated in their relationship to humans, but also that humans were domesticated in part because of their relationship to dogs. The reasoning is that the relationship between humans and dogs was a key catalyst for the advent of agricultural societies. To the extent that the rise of agricultural societies entailed the "domestication" of *homo sapiens*, dogs were integral agents in that domestication.[5] One might well be able to make a similar argument historically with regard to the human-falcon relationship, although that is mere speculation on my part.

While the arguments by evolutionary anthropologists about the co-evolution or co-domestication of humans and dogs are to some degree speculative and inferential, my argument does not rely on the veracity of their accounts. All that is needed is to call into serious question assumptions about the kind of life that is "natural" or "unnatural" for dogs.

Of course, none of these arguments imply that humans have in fact treated dogs well at all times or even most of the time. But it does problematize any claim that we inherently do wrong to dogs by keeping them as pets, working dogs, or companion animals. This I believe coheres fully with scriptural accounts not only of how our relationships with non-human animals *should go*, but also how they sadly *have gone*. While the scriptural narrative points us to the vision of peaceableness between all animals (including the human animal), it also recognizes the power of sin to deform these relationships.

From the beginning of Genesis to the end of Revelation, Scripture guides us in making meaning of who we are as human beings in relation to God, to other human beings, and to the rest of God's creation. Much of this "meaning making"—whether literal or symbolic—only goes on in the context of the relationship between humans and other animals. In the peaceable Edenic world of Genesis 1, all the animals, including the human

4 The morally problematic context for Belyaev's experiments are rarely if ever mentioned. In the nineteenth and twentieth centuries, silver fox fur was associated with royalty, and was highly sought after. Apparently, Belyaev started his experiment in 1959 at the behest of "fox farmers," who found that the silver foxes they were raising and slaughtering to be difficult to "work with." Belyaev's mandate was to come up with "friendlier" foxes to make the farmers' lives easier. So Belyaev bred foxes purely for the behavioral characteristic of being "friendly" rather than fearful or aggressive. Considering that Belyaev apparently only chose to breed the friendliest fox per hundred he tested, it would seem that the farmers would have had to provide Belyaev with thousands of foxes from which to find the few which he would breed.

5. Hare, "Survival of the Friendliest"; Shipman, "Do the Eyes have it."

animals, live at peace, with every animal eating only plants and vegetables. Furthermore, one of the two major tasks given to human beings in Genesis 1 is to name the other animals. "Naming" another is to give it significance, to come into a kind of relationship with the other. At least in Genesis 1, God seems to see humans relating peaceably to other animals as a crucial human responsibility. Furthermore, in the second creation story in Genesis 2:4—3:24, the Lord God envisages the other animals to be *the* companions for the man, and only after they are found to be inadequate companions (perhaps something to do with the other major command of Genesis to be fruitful and multiply) does God see fit to give Adam a human companion.

But not all is peace and harmony. It is a nonhuman animal that sets off the chain to the fall of creation, with the subtle serpent suggesting that disobeying the Lord will make human life far more "interesting." But it turns out to be a bad deal for all, with bad consequences for humans, for animals, and for the earth. Although the consequences for humans, for the serpent, and for Earth are laid out in Genesis 3, the rest of the nonhuman animals only discover their new fate in Genesis 9, where God says the other animals will now live in fear and dread of human beings. I take it that God is stating a fact, or at least offering a divine "prophecy," rather than issuing a command. And this fact—that animals will live in fear and dread of humans—has never been more true than in our present day. Whatever vaguely warm or friendly views we may think we may have towards "animals," we now live in a time where it has never been worse for the rest of God's animal creation. The sin committed with regard to factory-farmed animals in the last fifty to sixty years is undoubtedly greater than the sin against nonhuman animals in the rest of human history combined. And things are not much better for "wild" animals, as between deforestation and human-induced climate change, humans are undermining the kinds of habitat many species require in order to survive.

Max's Time with Us

After almost two years on the list for a failed therapy dog, we finally got a call about a dog we could adopt. We arrived at the facility to meet Billie, the most sickly two-year-old Labrador we could imagine. The poor dog was lethargic and completely indifferent to us. She was very small and underweight for her age. We were given a copy of her medical history and it was extraordinarily disheartening. Billie required insulin shots three times a day, was allergic to almost all dog foods (they were feeding her kangaroo meat), had a terrible skin condition, and had heart problems. Our hearts went out to this poor dog. But we knew that a family with three young children was not the right one for such a sickly dog. Furthermore, how could we bring home Billie to our son, knowing that she would likely die by the time she turned five?

This experience really got to us. Why could we not give a home to Billie? Were we only thinking of the perfect dog, one that would fit our expectations? Were we treating our potential dog merely as a commodity? I like to think not. If we had adopted Billie as a healthy dog and she then developed these health problems that would be one thing. She would already be a member of the family and we would have already committed to her. But voluntarily taking on this extraordinary commitment to nurse Billie, who was effectively a terminally ill juvenile dog, was another thing altogether. The moral life, as has been said, is not so much about taking on commitments, as living up to the commitments one takes on. We knew we neither could nor should take on that commitment, not for ourselves, and especially not for Jack.

Jennifer and I were now jolted into action. We had gotten Jack's hopes up about finally getting a dog, and now we turned it down. Anything less than a determined search would seriously harm our relationship with Jack, especially his ability to trust our promises. We had researched various breeds with him, and although Jen had an obsession with Great Danes, we had agreed we would adopt a Labrador Retriever. A family dog. Gentle. Loves water. Good with kids.

We looked on petfinder.com. We visited shelters and met a few companionable dogs, but each one was adopted by another family. We started to expand our search. Max showed up on the website. His short bio stated,

> *Max is a big love. He is going to be a big boy as he is a Lab/Great Dane mix. He is appropriately 8–9 months old, 75 lbs., and is Neutered, UTD, housebroken and crate trained. He is good with other dogs and children, cats unknown. He is a bit fearful of men, especially if they are wearing baseball caps, but we are working on that."*

Jennifer called Max's foster mom, an extraordinary lover of dogs who had rescued numerous abandoned and troubled dogs over the years, rehabilitating them, training and socializing them, and then "re-homing" them. His foster mom told us that Max had had a rough start in life. She had rescued him at six months of age, having found him leashed to a tree in the original owner's backyard, where he had likely been for much of the first six months of his life. He was a loving dog, but frightened of strangers, especially men. Max's foster family had three boys and two other large mature dogs, so Max began the process of being properly socialized by the other dogs as well as by his foster family. His foster mom told us she trained him to clearly recognize her as the boss. Max was doing well with basic training both on and off leash.

We felt we had met our dog! Meeting Max was complicated by the fact that he lived 300 miles away. Jennifer called Max's foster mom again to learn more about Max. His foster mom gushed about the friendliness and loyalty that Max had shown her family, stating, "He is going to make an awesome family dog. He will need to figure things out, but all in all, he only wants to be devoted to you and be one of your pack."

So our family of five piled in the car and headed off to meet Max. He seemed relatively obedient, affectionate, and loyal. We figured he would continue to make progress as he got older. He was certainly big—eighty pounds at nine months—and extraordinarily strong—and he captivated us. Even our seven-year-old daughter, who was generally nervous about dogs, gave him a tentative thumbs-up. We knew that if we wanted Max, we had to take him that day, as another family was eager to adopt him. So after an hour spent with Max we signed the papers to adopt him, loaded him up in our car, and made the trek to a vacation cabin on a lake. Max whimpered a bit on our car ride back, which worried Jack. When we arrived it was pitch black outside, which made the transition for settling down for the night challenging. Max was not too keen to go back in his crate. We put his crate in the boys' room at their insistence. But Max's whining and barking got to Jack, who entered our bedroom in tears, "This is so hard! I can't bear hearing him upset." We counseled him to reassure Max, but let him cry a bit. But Jack caved to Max's cries and spent the night sleeping on the floor in front of Max's crate.

In the beginning all was happy. Max loved being at the lake cabin, with swimming and space to roam. Extended family came to meet us and were impressed at what a well-behaved dog we had. After vacation we returned to Toronto, and Max started making friends at the local dog park. Many of the dog people at the park seemed surprised to hear how new he was to our family given his sociable, friendly, and well-behaved manner.

But life in the middle of a city proved more difficult for Max. He awoke each morning ready to spring from his crate. He was wound up and impatient to head to the park after his morning pee in the backyard. We were home for only two days before we again packed up the car to drop off our kids at camp for two weeks. Max accompanied us and excitedly followed our eldest son to his tent to drop off his bags. Scores of eleven-year-old boys flocked to Max's side to pet and play with him. Max tolerated the strange touches and rowdy behavior. Jennifer and I returned to Toronto childless for two weeks, but with our new canine family member.

During those two weeks we started a new rhythm of life. Max was quite obedient to me, at least when I was firm with him. He had incredible energy to burn off each morning, and he was strong. Jennifer, who had never had a pet, was less confident and decisive with Max, and lacked the strength to keep him in line. When she walked him he would lunge at squirrels, dragging Jen along in his pursuits. Alternatively, he would run towards the road, indifferent to passing cars. At the dog park he had a marvelous time, good-naturedly running and playing with the other dogs. Although he had been trained to respond to his name and usually would respond to my calls, at times he would ignore Jen when she called him, seemingly aware that he was being naughty. All these things upset Jen. What seemed like a power struggle was beginning.

We also quickly discovered that Max had a major protective streak. As we were now in close proximity to our neighbors, trouble was brewing. Our house shared an alley with our next-door neighbor, a friendly middle-aged gentleman who loved to garden. The first time they met, Max growled, pulled on his leash, and nipped our neighbor's shoe. Our neighbor was by nature wary of dogs, and Max had made a very bad first impression. He grew increasingly frightened of Max, despite our initial efforts to acclimate them to each other. Max sensed those who were fearful around him, and fearful responses seemed to confirm to Max that such persons were not to be trusted.

We hoped that in short order Max would become less wary of our neighbor. As long as Max had two to three walks per day, he seemed generally content and happy to be by our side. His favorite spot at the end of the day was to lie across our feet as we read on the couch.

Toward the end of our two weeks alone with Max, there occurred a devastating incident. A dear friend dropped by with her two young boys. They were all eager to meet Max. Jen and the three of them went off to the local park with Max. With no other dogs at the park, Max spent a little while futilely chasing a few squirrels, and then settled in the grass with a stick in his mouth. Jen got engrossed in conversation with her friend. The younger boy, who was six, bent right down to Max's level, and hoping to play fetch with Max, tried to take his stick away. Max leapt up to grab the stick back and nipped the boy. Apparently, Max put the boy's entire face in his mouth, as bite marks were visible in the vicinity of each of his ears. Thankfully, it was a nip, as a real bite could have literally defaced the boy. The poor boy was frightened and began screaming, "He bit me! He bit me! Am I going to die?" (Apparently the boy's father, to encourage him to respect dogs, had joked something to the effect that a dog bite would kill him.) While the wounds were shallow and quickly healed, we were devastated by the experience. Under Toronto's dog regulations, if both our neighbor and our friend had reported Max's bites, Max would have been permanently required to wear a muzzle whenever he left our house. Furthermore, as inexperienced dog owners, we found ourselves doubting our ability to take necessary care of Max and those he encountered.

The next day was Max's first check-up at the vet. Max growled at the vet more than once, and it was a struggle to get through the check-up. With regard to Max's biting the boy the previous day, the vet reassuringly stated: "That is a strict 'no, no' with a dog. Children should not take anything out of a dog's mouth." While mildly reassured that Max was not a peculiarly aggressive dog, it did not solve anything. In the two weeks we had had Max he had already bitten twice. Explanations and quasi-justifications did not ameliorate the existing problem. We enrolled in dog obedience classes, recognizing that what we really needed was "human training" classes. Jen felt that she needed to better understand the psychology and sociology of dogs in general and Max in particular. I was all too aware of innumerable dog psychology theories, which contradicted each other at key points. Jen noticed that Max responded very well to me, occasionally obeyed her and Jack, and merely tolerated our two younger children. In short order, we came up with lots of hypotheses. Was Max asserting his leadership (or "dominance")? Was he seeking to protect us in our home? Was he treating the rest of the family as peers, or as subordinates? Was he establishing himself in the "pecking order"? We had no idea which of these theories were bogus and which had some validity, but we were desperate to try to "figure Max out" so he could remain a member of our family. We noticed that Max at times attempted to herd the children when they were walking in the house, and later learned that Max periodically nipped the children when they were playing with their toys. When I was not around Jen and Jack would respond by trying to increase their air of authority, but it had relatively little effect. As we would find out soon, it was too little too late.

While we read some "respectable" guides to dog ownership and training, they were no substitute for experience and "presence" around a dog. For example, one afternoon when my octogenarian aunts came by, their genuine friendliness and complete lack of fear towards him preempted even the slightest growl. My septuagenarian uncle then followed my aunts through the front door. Max was more suspicious of males. "Uh-oh" I thought, and was poised to grab his collar. As my uncle crossed the threshold, Max immediately and slightly aggressively jumped up on him, landing his paws on my uncle's shoulder blades. I feared my uncle would topple backwards out the front door. But my uncle simply grabbed Max's front paws, laughed, went nose to nose with Max and spoke playfully to him. By the time my uncle got seated on the living room couch, Max was rolling around on the living room floor next to him, looking for belly rubs.

When I would notice people who seemed naturally fearful of Max, I would always think of Vicki Hearne's expression for such people—"natural bitees." With reference to walking one of her dogs, Hearne says: "I must be careful not to ask anyone who is a 'natural bitee' to approach and touch her. Natural bitees are people who . . . attempt to *infer* whether or not the dog will bite, jump up on them or whatever. . . . They are—sometimes only momentarily—incapable of beholding a dog. It is not that the required information will follow too slowly on their observations, but that they *never* come to have any knowledge *of* the dog, though they may come to have knowledge *that* . . . welcome has been withheld."[6]

It occurred to me that Hearne is pointing us towards what might be called "Job's task." In Job 39 the Lord tries to educate Job with regard to how little Job knows of God's ways, especially regarding God's relationship with God's other creatures.

> Do you know when mountain goats are born,
> or watch for the birth pangs of deer,
> Number the months that they must fulfill,
> or know when they give birth,
> When they crouch down and drop their young,
> when they deliver their progeny?
> Their offspring thrive and grow in the open,
> they leave and do not return.
> Who has given the wild donkey his freedom,
> and who has loosed the wild ass from bonds?
> I have made the wilderness his home
> and the salt flats his dwelling.
> He scoffs at the uproar of the city,
> hears no shouts of a driver.
> He ranges the mountains for pasture,
> and seeks out every patch of green.
> Will the wild ox consent to serve you,
> or pass the nights at your manger?
> Will you bind the wild ox with a rope in the furrow,
> and will he plow the valleys after you?
> Will you depend on him for his great strength
> and leave to him the fruits of your toil?

6. Hearne, *Adam's Task*, 59–60.

Can you rely on him to bring in your grain
  and gather in the yield of your threshing floor? . . .
Is it by your understanding that the hawk soars,
  that he spreads his wings toward the south?
Does the eagle fly up at your command
  to build his nest up high?
On a cliff he dwells and spends the night,
  on the spur of cliff or fortress.
From there he watches for his food;
  his eyes behold it afar off.[7]

Job's task is to learn to see in God's diverse creatures not primarily as creatures that might hurt him, scare him, help him, or serve as food for him, but rather first and foremost as creatures who have their own ends in God, and in whom God exults. Job's task is also our task, to overcome our presumption that all nonhuman animals exist as instruments for human use and enjoyment. While there are certainly a variety of indications in Scripture and Christian tradition that we must see the end of the rest of creation as primarily to glorify God, it is a major achievement of Pope Francis's *Laudato Sí* to unequivocally recognize and proclaim the intrinsic goodness of all nonhuman creatures.[8]

If, like Job, we too learn to rejoice in God's other creatures, our ingrained response will be one of wonder, seeing a reflection of God in them, a reflection upon which we want to meditate. Looking into the eyes of a snowy owl, or a wild boar, or a bottlenose dolphin, or an octopus, we aspire to a greater humility before the grandeur of God and God's precious creation. And so too, when we meet a new dog, what it means to see him or her as a reflection of God will be first and foremost to behold him or her, presuming conviviality and friendship. Sadly, sometimes the presumption of conviviality must be modified very quickly. The dog one beholds may be unable to respond in kind. The dog one beholds may have been bred and/or taught to distrust (some) humans, or there may be another pathology in the dog's constitution. But as Christians who see God's creation as fundamentally good, our presumptive reaction to encountering all creatures of God is to be one of wonder and conviviality.

7. Job 39; 1–12, 26–29.

8. In *Laudato Sí*, Pope Francis tells us that each and every animal his intrinsic value (§118), that each creature is "good in itself" (§140); that "other living beings have a value of their own in God's eyes" (§69); "that the ultimate purpose of other creatures is not to be found in us, [but] is God, in that transcendent fullness where the Risen Christ embraces and illumines all things" (§83); and finally that "the Risen Jesus is mysteriously holding [all the creatures of this world] to himself and directing them towards fullness as their end" (§100).

All that was said above was true of our encounter with Max. In the time we knew him it would be intense period of learning about the possibilities of a convivial relationship between Max and our family, our friends, and our neighbors. It would be a time to learn about the extent to which the evils Max had suffered early in his life could be be addressed and overcome. And ultimately, it would be a time to learn that which we could simply not know about Max, and how to respond in light of the ultimate mystery that Max, like all God's creatures, ultimately embodies.

Saying Goodbye to Max

One morning, about two months after Max's adoption, the kids were in the backyard with Max. It was another day of packing up the car for the lake cabin. Our next-door neighbor came walking down the shared alleyway between our two houses. Max immediately bolted from Jack towards our neighbor. Jack grabbed the neighbor's forearm in his jaws and dragged the neighbor away from our yard. Our neighbor had to forcefully shove Max off of him. The scene was frightening and chaotic for all involved. The children were upset, our neighbor was upset, and so was Max. Once again we dragged Max to his crate while Jen (a physician) attended to our neighbor's wound.

Once we were sure our neighbor was okay—though he was certainly not happy—we headed out to the cabin. Max loved it at the lake, but seemed more wary of our neighbors. We spent the next week hypervigilant as we watched Max interact with our kids and our neighbors. Max seemed hyper alert also, and would growl as our neighbors exited their own cabin. Max spent endless hours swimming in the lake and chasing loons, neither of which seemed to tire him out much. Inside the cottage he continued to scrutinize our younger children's activities and occasionally nip at them.

We called his foster mom to discuss the recent events. She reflected not only on Max's behavior, but on the events in our family and our own behavior. She gently informed us that, in her opinion, Max was reacting to being in a home where he was not sure exactly who was in charge. (A problem we had even before Max arrived!) Furthermore, she said, since Max did not see a strong leader in our family, he concluded that he was higher in the pecking order than some other family members. His foster mom assured us that Max's reaction to the neighbor was a sure sign that he had bonded to us as a family.

While her theory was interesting, and might possibly even had some truth to it, it did nothing to ameliorate what was now a very serious problem. We were deeply concerned about our neighbor's well-being, and knew he should not have to live in fear of being bitten by our dog. It also did not help that our neighbor happened to be a lawyer! Max was periodically nipping at our children, and now he was growling at our neighbors up at the lake. We had so hoped that Max could frolic in the extensive woods around the cabin, but now we had to be vigilant and keep him on a leash even there. Jack, who had for so long pined for a dog, and now was deeply attached to Max, was beginning to seriously doubt if he could handle him. Even if Max's foster mother theory was true, could we "fix" him in time?

My own sense was that the past abuse of Max—his being tied to a pole in the backyard for long periods of time in the early months of his life—had seriously undermined the kind of socialization that every dog—and every human for that matter—needs to learn in the early stages of their lives. His foster mother, through heroic effort, had started—but only started—the necessary socialization process. But she used pretty intense methods, and also had had the benefit of her two larger and older dogs to continuously keep Max in line.

We saw in Max—as we saw in ourselves—the brokenness of our world, the ongoing effects of sin and evil. If indeed Max's original human companions mistreated him, tying him to a pole for months as a juvenile, then we could expect tragic consequences from that. But how those consequences might be manifested, or indeed if they would be, we could not even hope to know. We speculated that considering his early months, he would distrust humans who did not behold him as Max, welcoming him in his unique individuality. Even with those who dwelt with him as Max, we concluded that he was not adequately socialized to entrust himself to more than a limited degree.

Jen and I decided to talk to Jack. How was he feeling about how things were going with Max? Jack dearly loved Max, but he also was not sure he could handle him, and was concerned about the way Max was acting around his younger brother and sister. What would happen to Max, Jack wanted to know, if we decided not to keep him? We told him that we would return Max to his foster mother, in fact that was in our adoption contract, but we first and foremost wanted to know how he would feel about it. Jack came to his own realization that returning Max was what we needed to do. But would we get another dog? We promised we would.

Jen called up the foster mother and told her that because of the various biting incidents, it was clear that our environment was not a good fit with Max. We asked to return him to her. She said she would be most happy to take him back, and find another family for him, but was clearly in some sense disappointed with us. I did not lose sleep over that, although it clearly stung Jennifer. However, Max's foster mother said she could not take him back for three weeks. If we were going to take Max back to his foster mother, we decided we should not take him back to Toronto, as we were not willing to take the chance that Max would bite our neighbor or someone else in that time. So when it was time to head back home, we found an excellent kennel in the country where Max would have lots of other dogs to play with.

Two weeks later, Jack would insist on doing the twelve-hour drive with me to go pick up Max at the kennel, take him to his foster mother, make the tearful goodbye, and then return home. On the trip, we took an extra day and visited the Baseball Hall of Fame.

Epilogue

When we got home Jen and I talked to Jack about getting another dog. Jack knew that in another four months, the family would be going overseas for six months as part of my sabbatical. We told him we could look for another dog now, but then we would have to be separated from our new dog for six months. Alternatively, we could make plans so we would get our new dog when we got back from the sabbatical trip. After the long wait, the experience with Billie the sickly Labrador, and then with Max, we were determined to do everything we could to make it work the next time. Previously, we had focused on a Labrador Retriever. But after the experience with Billie and Max, we decided to needed to go in another direction. I had for a long time loved the other famous breed from the home of the Labrador Retriever. We decided on a Newfoundland dog, a very big, incredibly gentle, slobbering water dog. Being a pescatarian family, we liked the idea of a dog who, if and when he needed animal protein, would likely be happy with fish. This time we would look to bond with our dog from an early age, and so decided on a puppy. We will get the dog from a reputable breeder, which means our Newfoundland will cost a small fortune. That brings up a whole other set of issues, but those are for another occasion.

Of course, Jen needed to know what would happen to Max, so she lurked on Facebook to keep up with Max's adventures. He would soon go to another family. But again, before long, Max was back with his foster mother. After that, he was adopted by an older couple who lived out in the country with a big spread. We were pretty sure that was the perfect fit. We got no more news about Max.

Bibliography

Francis. *Laudato Sí: On Care for Our Common Home*. Rome: Libreria Editrice Vaticana, 2015.

Gaita, Raimond. *The Philosopher's Dog: Friendships with Animals*. New York: Random House, 2002.

Haraway, Donna J. *When Species Meet*. Minneapolis: University of Minnesota Press, 2007.

Hare, Brian. "Survival of the Friendliest: Homo Sapiens Evolved Via Selection for Prosociality." *Annual Review of Psychology* 68 (2017) 155–86.

Hearne, Vicki. *Adam's Task: Calling Animals by Name*. New York: Vintage, 1986.

Shipman, Pat. "Do the Eyes have it?" *American Scientist* 100.3 (May–June 2012) 198–201.

On the Eyes of a Cat and the Curve of His Claw

Companion Animals as First Theology

—MATTHEW EATON—

Facing Feline Alterity

I remember the irreducible and irreplaceable alterity of my feline companion, Fargo. This cat incarnates the theology of this chapter as an in-spirational trace, or invisible co-creator, one whom I have followed in ethics and theology since our first meeting, which I here explore.

The eyes of this cat and the curve of his claw have, to quote Martin Buber, "the power to speak a great language," even if such a language is mute and transcends banal expression via signs and symbols, concepts and ideas.[1] This great language reverberates throughout my body as an ethical appeal, a mute expression of affect that is felt before it can be betrayed in language. The appeal calls my human sovereignty into question by exposing my finitude and once awakened by this exposure, I am converted to a divinity entangled within animality. The an-archy of this little feline voice demands I radically reimagine my approach to both ethics and theology, insofar as they have traditionally stood rooted in logocentric appeals to humanist reason, language, and experience.[2] Words betray what I want to

1. Buber, *I and Thou*, 75.

2. Humanist in this context refers to anthropocentrism, and should not be seen as conflicting with a general belief in the goodness and value of humanity. It references a way of approaching the world that sets up a certain view of the human, as language-bearing, self-reflexive, rational beings, as normative for judging the essence and value of others. For more, see Wolfe, *What is Posthumanism?* and *Animal Rites*; Derrida,

explore; yet, in wrestling with my encounter with feline eyes and claws, I might approximate the dialogue. As such, I begin by remembering the day I encountered this cat who would revolt against my sovereignty and teach me the meaning of an-archy.[3]

I am first exposed to Fargo when he is an eight-month-old kitten, himself exposed and vulnerable in a cage, awaiting adoption. From his cage, this cat expresses concern for his own being, appealing to me from an impoverished world by embodied resistance and (non)power that transcends humanist language and conceptualization. As I walk past this cage, and the cages of many other cats, I am under no particular compulsion to adopt a cat; I have other intentions today. Fargo, however, has neither interest in nor respect for my intentions. As I walk past these cats, most of whom nap lazily behind bars, Fargo stalks me. He notes, I soon learn, the skin of my nude arm, exposed, without anywhere to hide, and within his reach.[4] Unsatisfied with his impoverished world, as it was cramped and barren, his eye tracks my movement and he swipes at my flesh with extended claws. He catches me unaware as I walk by in the open; he wounds me at my point of exposure—my nude flesh is passive and powerless against to the curve of his claw. For a brief moment I am caught by this claw, held hostage, by this cat whose eyes gaze into mine, calling me into question for walking past him with no concern for his cry. With a paw, and a claw, he marks me, leaving behind a trace of himself that transcends time—a text written in blood red for me to read when I awaken within a world dominated not by bodies that feel, but minds that think, and speak, and construct concepts concerning

---

"*Geshlecht* II"; *Of Spirit*; *Aporias*; "Force of Law"; and *The Animal That Therefore I Am*.

3. The "an-archic" quality of ethical events is central in this essay, though I cannot fully explore the meaning of the idea. An-archy is an ethical temporality rooted in affects or feelings that take place prior to the ἀρχή (*archē*: beginning/origin/principle) of the awakened subjective consciousness of the thinking self. It is a matrix in which bodies feel out their worlds by means of sensuality, prior to being able to think and speak about such worlds within a subjective horizon. An-archy thus calls the sovereignty of subjectivity into question. For more, see Levinas, "Humanism and An-archy," 127–49; and *Otherwise than Being*, 99–129. For a deeper analysis of this issue in the context of relationships with otherwise-than-human animals, see Eaton, "Theology and An-archy"; and "Enfleshing Cosmos and Earth," 142–54. For an excellent exploration of religion and affect, see Schaefer, *Religious Affects*.

4. Nudity is a technical term that I explore in the following section. It reveals the vulnerability of my arm, literally exposed, with nothing between my skin and the curve of the feline claw. Yet, to be nude here is also to be caught within an affective matrix, without the clothing and protection of language, thought, and any inherited conceptualization of species difference. Prior to the clothing myself in thought, I am nude before the other animal, caught up in a space and a time where the power of language is neutralized, and unable to protect me from the mute appeal of alterity.

what has transpired in an absolute past. Like Kafka's jackals, he is a poor animal, having only a claw for everything he wishes to express and everything he wants to do; the only thing available to him is his claw.[5] The claw is not a hand; he is unable to grab me, hold me down, or utilize me for his freedom.[6] And yet the paw and the curve of the claw, as poor as it is, is still expression.[7] The only thing available to him is the power to wound exposed skin and thus rend time as the trace of this wound will be carried from the absolute past to the present where I might read his text and act responsibly. The claw, as I explain below, is the power of an-archy, disrupting my sovereignty, and creating me as responsible for his needs before I possess the freedom to decide whether or not I wish to act responsibly.

The animal who has caught me now faces me, looks me in the eyes, expressing in high-pitched shrieks of resistance, appealing directly, expressing that all is not well. Is this at all similar to what Buber once felt when he too gazed into the eyes of a cat and sensed a great language? In this relation, Buber senses a plurality of questions arise from the cat: "'Is it possible that you think of me? . . . Do I concern you? Do I exist in your sight?'"[8] I sense a similar expression from Fargo, albeit in a different mood. Fargo does not interrogate; his expression commands. "I know you're thinking of me; you're looking right at me! Clearly, I concern you and exist in your sight or else you wouldn't have stopped when I pleaded with you—you wouldn't have reached in here and touched my paw. This whole situation is your responsibility now; get me out of this cage!" Fargo pleads to neither live nor die alone; he pleads to escape his isolation. He inscribes this blood red text into my flesh, writing his body on mine with eyes and a claw.[9]

5. "We are poor animals, we have only our teeth for anything we want to do, good or bad, all we have is our teeth." Kafka, "Jackals and Arabs," 255.

6. On the idea of the hand as the linguistic power of naming, grasping, knowing, and controlling, especially as explored in the work of Martin Heidegger, see Derrida, "*Geshlecht* II" and *Of Spirit*, 47–57.

7. I here draw on and pervert ideas found in Levinas and Heidegger concerning animality and expression. Cf. Levinas, *Totality and Infinity*, 262. "The whole body—the hand or a curve of the shoulder—can express as the face." Cf. Martin Heidegger, *The Fundamental Concepts of Metaphysics*, 177. "We can formulate these distinctions in the following three theses: [1.] the stone (material object) is *worldless*; [2.] the animal is *poor in world*; [3.] man [sic.] is *world-forming*." Despite the radical differences, both Levinas and Heidegger are constrained by an anthropocentrism that limits their valuation of nonhuman animal bodies. Their thought, however, is ripe for perversion and reimagination in favor of the otherwise-than-human. Readers familiar with Levinas will note how dependent I am on his thought especially, which I embrace and pervert at each moment.

8. Buber, *I and Thou*, 75.

9. Cf. Cixous, "The Laugh of Medusa." Cixous, in an early text on feminism and

I come to, awakened by pain. I am confronted by a narrative of resistance inscribed in my body by another, a message from another time, the meaning of which I have felt before there is time to interpret. I understand the text because of an affective exegesis long past; I do not assess whether this animal has the neurological complexity needed to suffer as I do, nor determine whether he might be able to express anything like resistance in language. I do not calculate anything concerning the meaning of the encounter because there is no time for calculation; the meaning has been revealed prior to my freedom to interpret. Because I am a feeling body, open to concern for others prior to the choice of whether or not to be responsible, I awaken to a conversation that our bodies have already had and an agreement that we have come to, despite my subjective passivity. I awaken already responsible regardless of whether or not I choose to make good on this obligation, a choice to be settled in my present wakefulness. However, there is no denying that I awake as one responsible, having been converted by the irrefutable summons of this little cat who calls my sovereignty into question and demands a response.

Today, my arms are scarred from over a decade of encountering the curve of his claw. The claw and the glance of an eye express hunger, fear, and often joy—appeals in a plurality of forms to respect the other qua its alterity. The eye and the claw are expressions of an animal who overflows my thought and escapes my grasp. Such expressions, incarnate in the face of this cat, are beyond my ability for apprehension. They are beyond, counter-phenomenal, and disincarnate in their refusal to present the feline *as such* within my waking consciousness, and yet incarnate insofar as it is rooted in his embodied peculiarity and not a deeper, separate authority.[10] These counter-phenomenal expressions co-create and in-spire me as responsible, and from this dialogue the ideas of infinity, creation, and ethics open up, and God comes to mind.[11]

As such, I am forced to consider whether such encounters with animal alterity are not the (dis)incarnation of a God no longer enchained to a human face. Indeed, after encountering Fargo, the idea of a God separate

---

writing, describes writing as the writing of oneself as a woman in white ink. "Your body must be heard." Ibid., 880. While the parallels between feminist writing and animal writing do not synch absolutely, there are nevertheless similarities here and those ideas developed by Cixous.

10. The "disincarnation" of alterity refers to the idea that there is always more to the other than what any subject determines about its essence and identity. In this sense, alterity is infinite, or beyond what the subject is able to think concerning it. See Levinas, *Totality and Infinity*, 79; Gibbs, "The Disincarnation of the Word," 32–51.

11. On this idea of divine revelation, see e.g., Levinas, *Of God Who Comes to Mind* and "God and Philosophy," 153–73.

from all flesh no longer feels tenable. Divinity and materiality must exist in ontic unity, a religious ecology in which we live and move and have our being (Acts 17:28). Despite my inheritance of classical ideas concerning the structure of divinity, humanity, and animality, a door has been opened as a result of this encounter. On the other side of the door is a God revealed as otherwise-than-human and otherwise than the mediation of a phantom power belonging to a spirit divorced from all flesh. Revealed is a God who confronts me directly in animal flesh, gazing into my eyes and clawing at my sovereignty with a feline body. I see then neither difference nor separation between the vulnerable expression of animal bodies and the revelation of the divine command, "You shall not murder" (Exod 20:13).

Exposure and Nudity

At first glance, it may seem strange to describe Fargo's appeal as mute or resisting language as it is popularly conceived. He was, and remains, a loud cat, constantly demanding things of me with a plurality of vocalizations. The muteness of his appeal rests in the inability for the two of us to communicate via language, codified in signs, symbols, and concepts. The appeal erupts deeper than the common idea of language with its logocentric assumptions about what may or may not count as expression. His appeal is affective, erupting from the communicative depths of his body, the expression of eyes and claws, which my body feels.[12] Fargo and I commune in this affective matrix prior to and beyond language, by virtue of the ontic quality of our specific embodiment. In this sense, his appeal is mute, as our communication transcends language; the two of us are instead exposed to one

12. Affect in my usage, or at least what I specify as an-archic as opposed to awakened affect (see n. 14 below), refers to "the flow of forces through bodies outside of, prior to, or underneath language." Schaefer, *Religious Affects*, 4. I thus have in mind feelings outside of the power of a self-aware, conscious, intentionally minded subject. Affect here refers to the feelings erupting sensuously in bodies that precede and ground the possibility of intentional thought. I thus, with Schaefer, eschew the linguistic fallacy in religion, which suggests that religion is always an awakened, linguistic phenomenon. "The linguistic fallacy misunderstands religion as merely a byproduct of language, and misses the economies of affect—economies of pleasure, economies of rage and wonder, economies of sensation, of shame and dignity, of joy and sorrow, of community and hatred—that are the animal substance of religion and other forms of power." Ibid., 10. Religion, I insist, erupts from the an-archic affects that precede and make language possible, and that language describes what has already been expressed in one's body. On affect in Levinas, which I draw on below, see especially Levinas, "God and Philosophy"; Bergo, "The Face in Levinas"; and Peperzak, "Affective Theology, Theological Affectivity."

another, nude beyond the clothing of a conceptualist horizon restricted by concepts and logocentric ideas concerning language.

Exposure and nudity, key elements in the narrative recounted above, thus describe the mute appeal possible between humans and what is otherwise. Appeal and response occur beyond language, and beyond the present temporality wherein subjectivity resides and language, commonly understood, unfolds. Exposure and nudity concern the body prior to the linguistic and conceptualist constraints that arise when clothed within thought. When exposed and nude, without the covering of language, bodies are able to feel one another and respond to the sensuality of each insofar as their bodies allow. In the case described above, I become exposed to the exposure of a vulnerable cat.[13] His resistance to isolation appeals to me affectively as my body feels for his pain, rendering me responsible through embodied feelings that take hold of me prior to my own awakening as a subject.[14]

To more adequately flesh out exposure and nudity, I turn to a brief exploration of such ideas as they occur in the thought of Jacques Derrida and Emmanuel Levinas. In *The Animal That Therefore I am*, Derrida remembers the other animal who has been looking at him "since so long ago"—i.e., a temporality likely referring to the affective matrix prior to the awakening of

13. I understand vulnerability as a universal condition concerning an inability to be otherwise than finite. All bodies face the threat of mortality, decay, and being consumed, and thus transformed into energy by other beings. In the face of such finitude, beings resist that which frustrates their fulfillment. Such vulnerability points to an underlying quality of the cosmos, namely, that to be in relationship within a universe means to be open to the inevitability that your particular form of embodiment will not last, or that it will substantially change. Vulnerability cannot be compared across the boundary of different bodies, as each exists in the specificity of its differentiation. It is because of this that vulnerability cannot be normalized to occur only within specific modes of being, but is instead radically and infinitely plural. Such finitude solicits the human in affects such as sadness, empathy, and compassion that are irreducible to thought, and thus afflict the subject an-archically, asymmetrically, and infinitely, beyond the horizon of consciousness. The frailty of "not-being-able," as it erupts within human affects that lead to hospitality and care for the other, paradoxically becomes a dynamic, active resistance, what Levinas calls "the resistance of what has no resistance—the ethical resistance." Levinas, *Totality and Infinity*, 199.

14. Affects originate in a time prior to subjectivity, but erupt from the absolute past into the present. I identify these affects as an-archic or anoetic (i.e., the feelings and emotions that occur prior to self-reflective consciousness), and awakened or autonoetic affect (i.e., emotions subject to conceptual reflection as they travel from pre-reflexive sensuality to subjective consciousness). On this structure, see Panksepp, *The Archaeology of Mind*. Dialogue then may proceed through embodied feeling, beyond logocentric frameworks. Such expression is necessary to make sense of relations between different animal species, but need not be restricted to animality. For relations beyond animality, see Bennett, *Vibrant Matter*.

the subject.[15] For Derrida, the other animal who most clearly faces him and gives meaning to ethical temporality is a particular feline who calls him into question when exposed, and in this case, literally nude. Derrida's awakening to the other animal occurs after he has already been "caught naked, in silence, by the gaze of an animal . . . the eyes of a cat."[16] This gaze, for Derrida, is troubling. It leads to his recognition of embarrassment, vulnerability, and even a resistance to being exposed. Without anywhere to hide or anything to cover himself, Derrida is caught out in the open, face-to-face and vulnerable before the other animal. The shame of this encounter, arising from his exposure and nudity, and his shame at being ashamed, causes him to chase the cat away and retreat, as we will see, to the isolation and safety of his own horizon, being alone in a world apart from any relationship with the other animal.

Derrida's themes in describing his encounter with the cat—i.e., nudity, passivity, confrontation, isolation, and shame—are no doubt taken up in dialogue and frustration with Levinas, whose work sheds light on Derrida's encounter with the other animal even as it simultaneously refuses the depth Derrida assigns it.[17] While Derrida goes where Levinas will not, acknowledging a mutual exposure between himself and the otherwise-than-human animal, his remembrance of being "seen naked under the gaze of a cat," points to an ethical moment that Levinas develops and recounts more clearly throughout his writings.[18] For Levinas, the ethical event is wrapped up in the themes Derrida applies to the cat; ethics is an event of exposure and nudity, passivity and confrontation, isolation and shame that occur beyond the present temporality of subject. Thus, rather than proceeding from the

15. Derrida, *The Animal That Therefore I Am*, 3. Derrida here is drawing on Levinas's notion of "an-archy," the sensuous temporality where face-to-face encounters unfold prior to the awakening of a subject, a temporality of the present.

16. Ibid., 3–4.

17. Derrida's specific engagement with Levinas on ethics and the other animal, wherein he insists that Levinas's philosophy contains all of the necessary resources to embrace the otherwise-than-human, indicates that Derrida's ideas concerning the other animal cannot be understood apart from Levinas, despite his frustration with Levinas's anthropocentrism. Ibid., 105–18. I concur with Derrida's frustration with and perversion of Levinas's otherwise profound philosophy.

18. While I cannot delve into the complexities of Levinas's thought on the other animal, I note that even in his later thought he remains mostly closed to the otherwise-than-human. See Levinas et al., "The Paradox of Morality." Yet, the near consensus in modern philosophy, with some disagreement, follows Derrida in acknowledging that Levinas's thought, applied consistently, should be open to the other animal and provide a model to understand ethics between humans, other animals, and beyond. There are too many works on this topic for an exhaustive list. Two outstanding examples are Benso, *The Face of Things*; and Calarco, *Zoographies*.

constructions of a logocentric subject, a mind who assigns meaning to alterity by comparing it to this or that, ethics emerges prior to subjectivity, in face-to-face embodiment wherein "the one is exposed to the other as a skin is exposed to what wounds it, as a cheek is offered to the smiter."[19] To exist prior to thought and language is "a denuding, an exposure to being affected, a pure susceptiveness" to the other's corporeal expression before subjective conceptualization and intentionality.[20] As such, exposure and nudity for Derrida and Levinas speaks of one's life beyond the logocentrism of cognition, where affect expresses prior to and beyond subjectivity and awakens the I to responsibility and the freedom to embrace or deny the ethical summons. Thus, the one faced by the other is nude insofar as he or she is not yet a rational, language-bearing subject, but nevertheless a sensuous, feeling body. As Levinas writes, to be in this an-archic matrix prior to subjectivity and language is to be "without clothing, without a shell to protect oneself, stripped to the core as in an inspiration of air, an ab-solution to the *one*, the one without complexion."[21]

Yet, to be nude in this sense parallels nudity is a more literal sense, insofar as nudity might signify one's vulnerability. The nudity of the feeling subject becomes exposed to the literal exposure of the other, whose finitude and vulnerability solicit pathos in the subject. The nudity of the other thus refers not only to their existence prior to language, but to their suffering body, which may include literal nakedness—i.e., a body stripped from protection, vulnerable before the elements, a body that burns and shivers. This nudity is what finally expresses beyond any horizon, disrupting subjective powers by summoning responsibility by means of the affliction solicited in the feelings of the one it faces. The face expresses "in one's destitution and hunger."[22] Nudity then is life prior to and beyond the intentional cognitive horizon one the one hand, and corporeal vulnerability, perhaps literal nakedness, on the other. The subject is thus made responsible for another by means of affect, feelings for another's suffering erupting as one stands vulnerable, exposed, and nude before the vulnerable exposure of another's nudity. Ethics, then, is structured as this double exposure, a face-to-face

19. Levinas, *Otherwise than Being*, 49.

20. Ibid., 138. While not open to classical forms of epistemology, e.g., ontology, phenomenology, or empiricism, the existence of the face beyond the horizon of intentionality is always corporeal, even though such is not open to vision. "Sensibility can be a vulnerability, an exposedness to the other or a saying only because it is an enjoyment. The passivity of wounds, the 'hemorrhage' of the for-the-other, is a tearing away of the mouthful of bread from the mouth that tastes in full enjoyment." Ibid., 74.

21. Ibid., 49.

22. Levinas, *Totality and Infinity*, 200.

relationship that calls the sovereignty of the I into question through the pain that pleads for relief.

But what of Derrida's shame, and his being ashamed of being ashamed when exposed and nude before the other? What occurs in this space where one is in direct proximity to another without the protective clothing of a subjective horizon? Shame, for Levinas, is the awakened response to recognizing one's isolation as an ego alone in the world, enchained to itself, where every other is an alter ego, reductively identified by the subject without regard to their own transcendent voice.[23] Such an existential shame arises after one has been face to face with and called into question by another, a time and space where the two were in affective and bodily proximity. The shame arising after a confrontation with alterity uncovers and discovers an existence that is alone, isolated in a world where an awakened subject cannot help but reduce alterity to the themes constructed within a horizon.[24] The isolation of shame occurs insofar as one recognizes that there was once a time when two were in proximity to one another so long ago, in a time where responsibility emerged through pre-reflective affects. Such an ethical time, inaccessible now to the awakened subject clothed in thought, separates and isolates the I from the other, creating an unbridgeable gulf between the two. For Levinas, shame becomes the existential state of a subject stuck with itself, having been once been exposed to another in a time to which it cannot now return, a time in which the vulnerability of a suffering body pleaded for its well-being. Yet, and this is the entire point of Levinas's philosophy, subjects might now face their shame by embracing the responsibility to which one awakens, entering into an ethical relation with another and thereby escaping the chains of its isolation by caring for the needs of the one to whom they were once exposed.[25] When Derrida is ashamed of being ashamed, the ground of his embarrassment lays is his isolation, in distinguishing himself from the cat as one clothed in thought and language, and thus transcending the nudity of animal embodiment. And so, after being caught naked and exposed before this cat, Derrida recognizes his enchainment to his logocentric

23. "In the identity of the I, the identity of being reveals its nature as enchainment, for it appears in the form of suffering and invites us to escape. Thus, escape is the need to get out of oneself, that is, *to break the most radical and binding of chains, the fact that the I is oneself*." Levinas, *On Escape*, 55. Within a horizon, alterity has no voice of its own; it is identified as this or that, as determined by the subject.

24. "What shame uncovers [*découvre*] is the being who uncovers himself [*se découvre*]." Ibid., 64. See also Levinas, *Existence and Existents*, 84–100, where the analytic turns from shame to solitude, and Levinas introduces the other as a possible escape from existence.

25. Levinas, *Otherwise than Being*, 99–129.

self as the normative lens by which so often relate with the other animal.[26] In the an-archy of this encounter, this normative mode of human relationality is called into question and logocentric power is neutralized. He is thus not concerned with any banal impropriety before this cat; he is ashamed of his isolation from this "irreplaceable living being," whose mortality is his responsibility.[27] "Nudity," Derrida writes, "is nothing other than that passivity, the involuntary exhibition of the self. Nudity gets stripped to bare necessity only in that frontal exhibition, in that face-to-face."[28] There is then a different way of living, discovered in one's being uncovered by the gaze of another who solicits in a time beyond subjectivity, a time of an-archy where animal alterity calls human sovereignty and logocentric normativity into question and pleads for a life of peace beyond violence and theodicy.[29]

Ethics as First Theology[30]

Ethics is a denuding; it is the vulnerability of being stripped of logos and exposed to the disruptive affects grounded in the mute appeal of another's mortality. The responsibility created in such encounters is prior to one's

26. While there may in fact be truth in what is said about the other, such truth always betrays the pre-original saying in which the other expresses itself as such. "For the saying is both an affirmation and a retraction of the said. The reduction could not be effected simply by parenthesis which, on the contrary, are an effect of writing. It is the ethical interruption of essence that energizes the reduction." Levinas, *Otherwise than Being*, 44. Despite the truth of the said, the saying always overflows what is spoken. The pre-original saying refers to the affective encounter in which the passivity of the subject is exposed to the other, and made responsible prior to freedom. On the "saying" and the "said," see ibid., 5–7, 31–59.

27. Derrida, *The Animal That Therefore I Am*, 11.

28. Ibid.

29. For a critique of theodicy, which justifies violence, see Levinas, "Useless Suffering" and "Transcendence and Evil." Bringing theodicy into the discussion, if only for a moment and with little commentary, is important insofar as not all violence is avoidable. In the face of this lamentable fact, it is important to not justify unavoidable violence, but recognize that such acts of unavoidable violence will nevertheless be resisted, and thus can never be called Good. They are, at best, ambiguous aporias that allow some to flourish at the expense of others. They will allow a world to exist, but always be called into question. While literature on theodicy is legion, much more needs to be said in this area.

30. The idea of a first philosophy is classically connected to metaphysics derived from rational ideas and intentional thought, which can be traced as far back as Aristotle and later in the Enlightenment reimagined in the work of Descartes. Ethics, however, following Levinas, is prior to and the ground of all subsequent philosophy and theology and thus take priority over the rationalism and intentionality of metaphysics.

freedom to choose whether or not to be responsible. The wounded subject awakens as one already responsible for the other's suffering and eventual death. The summons to responsibility, taking place in a past in which the subject was never present for a life that looms on the future horizon, is revealed in as the divine command, "You shall not murder" (Exod 20:13). The event of ethics, in its an-archic temporality, grounds the possibility of awakening to thought and serves as first philosophy, but it also grounds the revelation and idea of God, who comes to mind in the face of the other.[31] "There are these two strange things in the face," Levinas writes, "its extreme frailty—the fact of being without means and, on the other hand, there is authority. It is as if God spoke through the face."[32] Thus, "an original ethical event . . . would also be first theology."[33] So, I suggest, following Levinas and Derrida, that theology is properly a reflection on ethics, made possible by affective encounters with vulnerable bodies who solicit pathos in subjects who awaken to the disruptive power of an-archy.

Thus, insofar as an individual is called into question by the ethical authority of another who transcends the subjective horizon and in-spires the subject as being-responsible, that person is confronted with the moral power, creativity, and infinite transcendence classically reserved for God. While Levinas reserves such divine expression for the human, my own confrontation with the power of transgressive, animal bodies calls this anthropocentric bias into question. Bodies, human or otherwise, possess the power to express themselves, thus in-spiring or co-creating differentiated subjects by shaping the way they relate to their world, even to the point of soliciting humans to ethical becoming in a manner typically reserved for divinity. This calls into question not only the sovereignty of the Western subject, and the nature of the subject/object relation, but the dichotomous relation between divinity and corporeity inherited in the Western philosophical tradition. Ethical authority, creative power, and transcendent being thus lie squarely in individual bodies, and not a greater alterity mediated by that which faces us. The God who comes to mind in face-to-face relations is not finally separable from corporeity itself.

31. See Levinas, *Of God Who Comes to Mind* and "God and Philosophy." The other is not precisely God in any reductionistic sense. Yet, the face is divine. The divinity of the individual seems to escape particularity, erupting within an infinite plurality of faces. As such, Levinas affirms the divinity of the other person, while denying that one's neighbor is God as such. In a radical kenosis, the divinity of each individual transcends its particularity, refusing any strict identification in any singular being. God is both particular and radically plural in this framework.

32. Levinas et al., "The Paradox of Morality," 169–70.

33. Levinas and Robbins, *Is it Righteous to Be?*, 182. For a nuanced analysis of Levinas's understanding of theology, see Cohen, *Levinasian Meditations*, 296–313.

Theology, I suggest, arises out of the infinitely complex matrix of affective human relationships with the world, integrating humanity, animality, and otherwise. It unfolds as a disruptive, an-archic memory, which opens up the idea of ethics, creation, and transcendent infinity as the ground out of which God comes to mind. As an-archic remembrance, theology recalls an affective, ethical encounter to which we were never present, and yet an encounter to which we were passively exposed, as skin is exposed to that which wounds it. Theology then is a reflection within a horizon, manifest in speech, but what is said in theology always betrays a pre-original event, an embodied "saying" expressed in the face of another to whom we were once exposed—in my case, a cat named Fargo. After meeting this cat, I am unable to think of theology otherwise. His ethical appeal—expressed in his pleading eyes and the curve of his claw—erupts as first theology. He calls my human sovereignty into question; his is "a voice that commands: an order addressed to me, not to remain indifferent to that death, not to let the other die alone."[34] Such is only the voice of a little cat, but the voice is divine.

Bibliography

Bennett, Jane. *Vibrant Matter: A Political Ecology of Things*. Durham, NC: Duke University Press, 2010.

Benso, Silvia. *The Face of Things: A Different Side of Ethics*. Albany, NY: State University of New York Press, 2000.

Bergo, Bettina. "The Face in Levinas: Toward a Phenomenology of Substitution." *Angelaki* 16.1 (March 2011) 17–39.

Buber, Martin. *I and Thou*. London: Continuum, 2008.

Calarco, Matthew. *Zoographies: The Question of the Animal from Heidegger to Derrida*. New York: Columbia University Press, 2008.

Cixous, Hélène. "The Laugh of Medusa." *Signs: Journal of Women in Culture and Society* 1.4 (1976) 875–93.

Cohen, Richard. *Levinasian Meditations: Ethics, Philosophy, and Religion*. Pittsburgh: Duquesne University Press, 2010.

Derrida, Jacques. *The Animal That Therefore I am*. Edited by Marie-Louise Mallet. Translated by David Wills. New York: Fordham University Press, 2008.

———. *Aporias*. Stanford, CA: Stanford University Press, 1993.

———. *The Beast and the Sovereign*. Edited by Michel Lisse, Marie-Louise Mallet, and Ginette Michaud. Translated by Geoffrey Bennington. Chicago: University of Chicago Press, 2009.

———. "*Geshlecht* II: Heidegger's Hand." In *Deconstruction and Philosophy*, edited by John Sallis, 161–96. Chicago: University of Chicago Press, 1987.

34. Levinas, "Diachrony and Representation," 169.

———. *Of Spirit*. Translated by Geoff Bennington and Rachel Bowlby. Chicago: University of Chicago Press, 1989.

———. "Force of Law: The Mystical Foundation of Authority." In *Deconstruction and the Possibility of Justice*, edited by Drucilla Cornell, Michael Rosenfeld, and David Gray Carlson, 230–99. New York: Routledge, 1998.

Eaton, Matthew. "Enfleshing Cosmos and Earth: An Ecological Christology of Deep Incarnation in Dialogue with Emmanuel Levinas' Ethics of Infinity." PhD diss., University of St. Michael's College, 2016.

———. "Theology and An-archy: Deep Incarnation Christology Following Emmanuel Lévinas and the New Materialism." *Toronto Journal of Theology* 32.1 (2016) 3–15.

Gibbs, Robert. "The Disincarnation of the Word: The Trace of God in Reading Scripture." In *The Exorbitant: Emmanuel Levinas between Jews and Christians*, edited by Kevin Hart and Michael Alan Signer, 32–51. New York: Fordham University Press, 2010.

Kafka, Franz. "Jackals and Arabs." In *The Metamorphosis, In the Penal Colony, and Other Stories: With Two New Stories*, translated by Joachim Neugroschel, 252–57. New York: Scribner, 2000.

Heidegger, Martin. *The Fundamental Concepts of Metaphysics: World, Finitude, Solitude*. Translated by William McNeill and Nicholas Walker. Bloomington, IN: Indiana University Press, 1995.

Levinas, Emmanuel. "Diachrony and Representation." In *Time and the Other and Additional Essays*, translated by Richard A. Cohen, 99–129. Pittsburgh: Duquesne University Press, 1987.

———. *On Escape: De L'évasion*. Translated by Bettina Bergo. Stanford, CA: Stanford University Press, 2003.

———. *Existence and Existents*. Translated by Alphonso Lingis. Pittsburgh: Duquesne University Press, 2001.

———. "God and Philosophy." In *Collected Philosophical Papers*, translated by Alphonso Lingis, 153–73. Dordrecht, The Netherlands: Nijhoff, 1987.

———. "Humanism and An-archy." In *Collected Philosophical Papers*, translated by Alphonso Lingis, 127–49. Dordrecht, The Netherlands: Nijhoff, 1987.

———. *Of God Who Comes to Mind*. Translated by Bettina Bergo. Stanford: Stanford University Press, 1998.

———. *Otherwise than Being or Beyond Essence*. Translated by Alphonso Lingis. Pittsburgh: Duqusene University Press, 1998.

———. *Totality and Infinity: An Essay on Exteriority*. Translated by Alphonso Lingis. Pittsburgh: Duquesne University Press, 2013.

———. "Transcendence and Evil." In *Collected Philosophical Papers*, translated by Alphonso Lingis, 175–86. Dordrecht, The Netherlands: Nijhoff, 1987.

———. "Useless Suffering." In *Entre Nous: On Thinking-of-the-Other*, translated by Michael Smith and Barbara Harshav, 91–101. New York: Columbia University Press, 1998.

Levinas, Emmanuel, and Jill Robbins. *Is it Righteous to Be?: Interviews with Emmanuel Levinas*. Stanford, CA: Stanford University Press, 2001.

Levinas, Emmanuel, Tamara Wright, Peter Hughes, and Alison Ainley. "The Paradox of Morality." In *The Provocation of Levinas: Rethinking the Other*, edited by Robert Bernasconi and David Wood, 168–80. London: Routledge, 1988.

Panksepp, Jaak. *The Archaeology of Mind*. New York: W. W. Norton, 2012.

Peperzak, Adriaan. "Affective Theology, Theological Affectivity." In *Religious Experience and the End of Metaphysics*, edited by Jeffrey Bloechl, 94–105. Bloomington, IN: Indiana University Press, 2003.

Schaefer, Donovan. *Religious Affects: Animality, Evolution, and Power*. Durham, NC: Duke University Press, 2015.

Wolfe, Cary. *Animal Rites: American Culture, the Discourse of Species, and Posthumanist Theory*. Chicago: University of Chicago Press, 2003.

———. *What is Posthumanism?* Minneapolis: University of Minnesota Press, 2010.

Animals as Eschatology

Struggle, Communion, and the Relational Task of Theology

—TIMOTHY HARVIE—

[I]t is clear that without touch it is impossible for an animal to be.

—ARISTOTLE, *DE ANIMA*, 434B 22–23

[T]he human race, the universe itself [are] closely related. . . . The promised and hoped for restoration, therefore, has already begun . . . and the renewal of the world is irrevocably under way.

—*LUMEN GENTIUM*, 48

BONES

Canines and humans have been relationally and communally entangled for tens of thousands of years. Debates on the origins of human-dog relations vary between hunter-gatherer narratives and the slow domestication of wolves[1] and accounts of genetically disposed canids becoming incorporated into human communities by scavenging through the waste and rubbish nearby early human communities.[2] Because of this long history and the present ubiquity of dogs in human homes, canine com-

1. Hobgood-Oster, *A Dog's History of the World*, 5–32.
2. Coppinger and Coppinger, *Dogs*; *What is a Dog?*

panions are considered by many to be family members and friends, ideally suited for life among humans. This view is so engrained it is understandable that it is often considered a given that when one welcomes a young puppy into one's home that the intimate relationship is immediate. A colleague once said to me that because of this she considered dogs too dependent, too subservient, even "too slavish" in temperament and so were less palatable as household companions than other species. The sentiment expressed appeared to be that dogs are too eager to please and lack an independence of mind and temperament. Such was not my experience.

Unlike some who write on their experience with dogs,[3] I did not grow up in a home where other animals were present. Furthermore, living in suburban Toronto, Calgary, or Regina in Canada, and briefly in Houston in the United States, regular exposure to other species of animals in the wild was not a part of my upbringing. When, in 2013, my future spouse and I were discussing getting a dog, I was hesitant but not closed to the idea. The primary responsibility for care would be mine until we were married, but I was assured by various people that a small dog would add joy and companionship to my home. Discussions continued for some weeks as we researched breeds that might be conducive to our lifestyle, health, and personal temperaments. I was intent on ensuring that this would be the puppy's home as it was mine. After these deliberations we decided to adopt a dog.

Soon after, we discovered people with adoptable puppies born from a recent litter. Driving to a town just north of the city where we lived, we arrived to pick up the dog. The inside of the house was unkempt and lacked the order and cleanliness I envisaged as important for rearing puppies. The floors were covered in blankets and strewn with random objects and children's toys. The children of the home were approximately five- and ten-years-old respectively and were being rambunctious as children can be. The tables were covered in papers and there was food in various places on the counters and the floor. The Bichon Shitzu mutt we found here had dubious parentage and was the last of his group as no one had sought him out. He was nearly twelve weeks old and past the age when much essential socialization occurs for a pup.[4] I did not yet realize how important those early weeks were and was naïve to the type of situation in which the puppy had first experienced life. He appeared as more a small sphere of fur than anything else, but he was not averse to us holding him and so the decision was made to take him home. We named him Bones.

3. Gita, *The Philosopher's Dog*.

4. Monks of New Skete, *The Art of Raising a Puppy*, 40–60.

Bones's entrance into my home was akin to welcoming a tornado into a small, confined space. He liked to play and did not like it when, after hours, I wanted a break from the repetitive task of throwing his favorite stuffed toy across the room so that he could fetch it, circle the coffee table twice with it, and then, when the mood struck, return it to me. These were the joy-filled moments one expects with a puppy. However, I quickly learned that Bones had some problems as well. It appeared he had not been well cared for in his first weeks of life and he required two surgeries within the first ten days of our shared life. The cause of these surgeries was fairly serious and the excellent veterinarian caring for Bones indicated he might experience the effects of the problem throughout his life, particularly as he grew older. In addition to the obstacles with his physical health, there were emotional and psychological hurdles that began to emerge in various behaviors.

Bones exhibited profound claustrophobia in any confined spaces. If he was placed in a kennel for safe travel to visit another dog, go to the vet clinic, or in a misguided attempt to make such a space his bed, he would panic. It is difficult to accurately describe his response. He did not bark or wail. Rather, the sound he made was more akin to a scream, accompanied with a clawing that put him at risk of injuring himself. My spouse and I realized quickly that the kennel ought to be put away permanently so that Bones would never see the object of his panic again. In addition to the fear of confined spaces, Bones exhibited a highly selective fear of children. When going for walks in the local neighborhood, there would often be children playing or walking nearby. If a child was a toddler, aged approximately between two and four years of age, Bones responded well. If the child appeared to be approximately twelve years of age or older, Bones was equally amiable and open to interaction. However, if the child appeared to be between the ages of approximately five and ten years of age he would suddenly cower in fear behind me, using my body as a shield. He would shake for minutes after an encounter and would often urinate uncontrollably if such children drew near. The significance of the apparent age of the children he feared was not lost on me and likely indicated mistreatment in his initial home.

Bones also exhibited anxious behaviors in the home that I was not well-equipped to handle. Like many dogs, and puppies in particular, Bones liked to chew. However, Bones would chew and bite everything. It is a common metaphor to say that one might be eaten out of house and home. Bones did not grasp the concept of metaphor. Despite having multiple toys of varying size, shape, and texture, Bones's destructive chewing was a severe manifestation of his anxiety. This small dog would habitually focus on one or two areas of the house and proceed to gnaw incessantly until either much damage was done or he was removed. He ate the corners of walls, decimating the material that provided a stairwell entranceway its cohesion, leaving only scattered piles of dust and drywall as evidence. He showed similar distain for stair edges regardless of whether they were wood or tile. Bones also offered me the opportunity to replace the sofa in the main sitting room after he managed to tear and eat entire portions of its cushions.

Many dogs chew in their puppy years. Bones obsessed. His behavior appeared more of a coping mechanism and a manifestation of internal lack of ease rather than puppy learning experiences. Bones would also respond anxiously if I intervened in one of these obsessive moments. He would tear my clothing, occasionally my flesh, and blood was drawn in more than one instance. I often sensed my own patience for this small creature waning. I fruitlessly attempted to train and discipline the dog over the first six months of our life together. I would attempt to incentivize certain behaviors through rewards or have Bones expend his stored energy through multiple long walks or running free at off-leash parks. I would attempt to curb his behaviors and he would continue to struggle with anxiety. This behavioral anxiety was likely due to less-than-ideal circumstances in the key formative weeks at the start of his life. Bones was brought into life in an unclean, neglected environment without essential formative relationships of nurturing and care and this had a large impact on him. My own reactions and inability to understand and provide the type of responsive, knowledgeable care that he required led me to approach him in objective ways. I sought to govern his actions rather than enter into a relationship. This was a daily cycle of stress, anxiety, and struggle in which both Bones and I participated. We treated each other as objects which lacked understanding and lived among each other rather than be truly present to each other.

This changed approximately six months into our life together when Bones initiated a new manner of relating. After a particularly challenging day I came home after working for the afternoon and took Bones outside as per our usual routine. Eventually, I reclined on a partially consumed sofa to rest but tensed as the small dog approached. Rather than chewing, scratching, or biting, Bones did something unexpected. He did something he had never done before. He leaped up onto my reclined torso and lay down on me. He gently licked my folded hands once and placed his head on my chest. I was aghast and did not dare move and risk disturbing him. To the best of my knowledge this action had never been modeled to Bones and I had certainly never seen him do this. We lay there for the next hour feeling each other's lungs expand and contract, sensing the rhythmic pulsing of one another's heart pumping blood throughout our bodies, and my sensing the warmth of the higher internal body temperature dogs have when compared to humans. This seemed like the first time Bones felt genuinely safe with me and I with him. This embodied, affective exchange of mutual vulnerability and presence communicated something new between us. For the first time, we engaged each other as subjects to be present with rather than objects to make demands of. Bones initiated this of his own accord. Julia Schlosser has commented that an "unconstrained animal is free to touch back or not by her choice, and a touch provides both participants with a reciprocal exchange of visceral or embodied information about the other . . . a felt-sense that often functions outside the patterns of language."[5]

This haptic communication through embodied presence immediately became a daily ritual. Each day I would recline on the couch and Bones would come, unsolicited, and lay upon me. This ritual became a liturgy and these liturgical movements structured our afternoons and evenings together and created a relational space between our bodies that communicated presence, vulnerability, and companionship. The daily practice provided order to the time we spent together. In developing the shared experience of this liturgy of embodied presence, Bones initiated and forged a friendship with me that I was blessed to reciprocate. Bones continues to have some struggles with anxiety, albeit not in the same destructive forms as he once did. However, this is recontextualized in a relational matrix that originated and receives ongoing impetus through our embodied encounters of kinesthetic presence.

Touch

Aristotle argued that flesh is the medium of touch. He wrongly considered flesh to be a membrane, that is a medium, rather than the organ of touch, but he recognized how essential touch was for all animals. While Aristotle did not think that other animals had "true flesh" in the manner of humans[6] he did argue that touch is the foundational sense necessary for animal life. "The primary form of sense is touch, which belongs to all animals."[7] Touch allows an animal, human and other, to interact with the world, locate itself within its environment and discover food necessary for survival. While taste is a unique sense meriting its own discussion in Aristotle's work on the soul, taste cannot exist without the sense of touch. Touch is a necessary prerequi-

5. Schlosser, "Tangible Affiliations," 31. In this action Bones overcomes the failure of what Donovan Schaefer has called the "linguistic fallacy." Schaefer, *Religious Affects*, 4–10.

6. Aristotle, *De Anima*, 423a 14–16.

7. Ibid., 413b 3–4.

site for taste, since one cannot taste an object without contact, and because taste is necessary for sustenance, touch is foundational for all animal life. This differentiates touch from other senses such as sight and sound. Touch requires proximity to an object or another subject. It requires one to come close to another and to interact with the other. Aristotle notes this feature of the body when he states, "for some community of nature is presupposed by the fact that the one acts and the other is acted upon, the one moves and the other is moved; interaction always implies a special nature in the two interagents."[8]

When touch is used as a sense of proximity I detect the closeness of an object to me. I sense the presence of a wall, a table, or a misplaced toy beneath my feet. In this way touch is a sense essential in the process of human learning. In these embodied encounters touch becomes a sense facilitating an increase of knowledge. "Touch gives animals the sense of inquisitiveness and builds the foundation of knowing."[9] However, touch is also a sense of perception. It not only allows one to know the presence of an object but its specific contours as well. The embodied interaction of the flesh of the animal with the physical being of an object allows the animal to understand whether this locale is safe or threatening, whether it is a place to be sustained and nurtured or whether the animal must move to a new place. Aristotle argues that touch is indispensable for this animal function that allows for all sensate life. "[E]very body is tangible, perceptible by touch; hence necessarily, if an animal is to survive, its body must have tactile sensation."[10] So essential is touch that Aristotle's medieval commentator, Thomas Aquinas, argues that it is convertible with animality itself.[11]

When touch is used as perception we are set in a particular, interactive relationship with another. David Abram writes that "the event of perception unfolds as a reciprocal exchange between the living body and the animate world that surrounds it."[12] This interaction requires vulnerability as it moves beyond unidirectional acquisition of knowledge where an object is observed, analyzed, and understood. It entails that we expose ourselves to the specificity of the being of the other. We recognize something of our self as well as the other. "If we notice something, we notice also that it is we who notice it. We recognize this, not only with the mind alone but also by means of some sense. . . . Today we call this psychical function consciousness. Thus,

8. Ibid., 407b 15–20.

9. Bremer, "Aristotle On Touch," 80.

10. Aristotle, *De Anima*, 434b 10–15.

11. Aquinas, *Aristotle's De Anima*, 496.

12. Abram, *The Spell of the Sensuous*, 73.

the act of perceiving is a synthesis in which the subject and the object are reduced to an indivisible synthesis."[13] To recognize oneself with another is to recognize oneself as a perceiver; that is, to recognize oneself through embodied touch. It is through touch that one is connected to another being and opens a space between two living animals. This mode of embodied relationality is revelatory by virtue of perceiving the other being as distinct from yet connected to oneself. This revelation offers new modes of being, new ways of being in the world, and even recreates the manner of one's own being in the world.

Edith Wyschogrod identifies this type of touch with empathy. Wyschogrod defines empathy as feeling-acts that "are structures of intersubjective encounter."[14] "Empathy" Wyschogrod writes, "[is] the feeling-act through which a self grasps the affective act of another through an affective act of its own."[15] The kinesthetic communication of the self occurs in the embodied, tactile modes of one animal with another. In this sense, to touch is always simultaneously to be touched. "The relationship between myself and other(s) and to the shared object, event, or value are equiprimordial and experienced simultaneously."[16] This occurs when Bones moves his body to rest across mine. The haptic fields of our sensory flesh make both of us vulnerable in this moment. We are exposed as canine makes contact with human. Our self-protective intuitions of biting and correcting, chewing and discipline are set aside in favor of mutual communion with the other as we are in that moment. It involved the risk of forging new modes of relating to each other in a manner that requires trust. In the liturgical movements of our shared, creaturely life the vulnerable exchange of desire for friendship creates new avenues for divinity to be revealed.

Martin Buber describes a similar exchange in a well-known passage where he recalls a boyhood experience with a horse.

> When I was eleven years of age, spending the summer on my grandparents' estate, I used, as often as I could do it unobserved, to steal into the stable and gently stroke the neck of my darling, a broad dapple-grey horse. . . . If I am to explain it now, beginning from the still very fresh memory of my hand, I must say that what I experienced in touch with the animal was the Other, the immense otherness of the Other, which, however, did not remain strange . . . but rather let me draw near and touch it . . . it

13. Bremer, "Aristotle On Touch," 77.

14. Wyschogrod, "Empathy and Sympathy as Tactile Encounter," 27.

15. Ibid., 27–28.

16. Ibid., 40.

> let me approach, confided itself to me, placed itself elementally in the relation of Thou and Thou with me.[17]

Here Buber gives expression to a tactile liturgy he experiences with the horse. It is through touch that he states "the element of vitality itself bordered on [his] skin."[18] This repeated sense of touch becomes more than a physical act; it embodies the creation of new modes of being and the possibility for the divine to enter into the mundane, natural reality of his existence. The rhythmic pattern of touch structures time and thus serves as the eruption of divinity into such relationships. In Buber's experience with the horse, what is witnessed is a haptic gesture that is more than the sum of its parts. Like Buber's horse, the liturgy Bones presided over offered a contemplative approach to experience God in our shared creatureliness. It is a complex exchange enacted in a single movement. Meaning is not merely conveyed, but generated in the mutual presence explored in the vulnerability of such feeling-acts.

Buber's own views pertaining to the inclusion of the nonhuman into the potential of intersubjective dialogue is complex and, at times, inconsistent. In his seminal early work, *I and Thou*, Buber reflects variably on how the nature of a tree or the eyes of an animal have the potential for mutual, intersubjective dialogue.[19] However, in the 1957 afterword to this work Buber argues that it is humans that move toward animals to welcome them in and make animals compliant to human life. He uses the metaphor of a threshold. He argues that animals are at the threshold of mutual relation, but humans are at the over-threshold. There is some reciprocity here in the afterword, but it is a graduated reciprocity where only humans have full capacity for true mutuality.[20] Peter Atterton notes this when he states, "the I-Thou relationship with the human other enjoys a certain privilege in Buber's descriptions. . . . But it is clear that the regions of nature—from rocks to plants to animals—are still defined in terms of their capacity for mutuality."[21]

In an article, originally published in 1950 under the title "*Urdistanz und Beziehung*," Buber reclassifies his earlier twofold language of I-It and I-You into a twofold movement of human life. Here Buber argues that the twofold movement is: 1) a primal setting at a distance, and 2) entering into

17. Buber, *Between Man and Man*, 26–27.

18. Ibid., 27.

19. Buber, *I and Thou*, 58, 144–45.

20. Ibid., 172–73.

21. Atterton, "Face-to-Face With the Other Animal?," 264. See also, Garrett, "Extreme Humanism."

relation (*In-Beziehungtreten*).[22] In this reconfigured framework, the first movement is the presupposition of the second, though it does not necessarily lead to it. Buber states that humans "can set at a distance without coming into real relation with what has been set at a distance."[23] In this essay, the elevation of the human in its ability to overcome and manifest relation becomes apparent. "An animal never succeeds in unraveling its companions from the knot of their common life. . . . Man and many animals have this in common, that they call out [*Anrufen*] to others; to speak [*Anreden*] to others is something essentially human."[24] In this essay only humans have what he calls synthesizing apperception, the apperception of being as a whole and a unity. In his more analytical moments, Buber responds to the question of mutual reciprocity between humans and animals with an affirmative tone, but a highly qualified one. This contrasts starkly with his own narration of his experience with the horse.

Buber describes the initiative of the horse. The horse welcomes him and allows him to draw near. In the simple gesture of human hand placed upon equine mane the divine enters into what Buber identifies as the "between" (*Zwischen*). In this space of the between, manifested in the tactile impression of touch, "the dialogical must become transcendental."[25] The simple unity of corporeal contact belies the complex, transformative presence of divinity as God enters this space. This complexity in simplicity is present in Bones's holophrastic gesture in commencing our liturgy of touch. It is in such moments that Buber identifies as full of potential for divine indwelling (*Einwohnung*) within creation. It is here that "the eschatological conception breaks into the lived hour and permeates it."[26]

Eschatology

"Communicative meaning" writes Abram "is always, in its depths, affective; it remains rooted in the sensual dimension of experience, born of the body's native capacity to resonate with the other bodies and with the landscape as a whole."[27] The account of touch developed by Aristotle indicates that this

22. Buber, *The Martin Buber Reader*, 207.

23. Ibid., 209.

24. Ibid., 210.

25. Tallon, "Affection and the Transcendental Dialogical Personalism of Buber and Levinas," 52.

26. Buber, *The Martin Buber Reader*, 91.

27. Abram, *The Spell of the Sensuous*, 74–75.

perception contextualizes all animal life in a relational matrix that can be understood in empathetic, revelatory terms. Buber opens the potential of such moments to be understood as having the capacity for not only creaturely communication, but in the relational between-space of mutuality the divine indwelling of eschatological presence. In the liturgical practice initiated in Bones's movement to inhabit space with me we are both changed in what occurs in the between. This ritual of touch becomes redemptive for the both of us. Bones begins to heal of much of his anxiety and begins to live with an air of joy and engaged attentiveness that comes with friendship. For my part, I am altered in how I perceive myself in the world, as but one member in a natural communion of animals and world as we inhabit creation in ways unique to our species-specific flourishing and eschatological hope of renewal.

Such a vision for eschatology runs counter to recent attempts to articulate an eschatological vision that emphasize conjecture of future states at the expense of divine immanence in the life of the kingdom manifested in the present. These proposals tend to emphasize the importance of speculative endeavors for doctrinally coherent arguments pertaining to the future destiny of creation and human participation in the beatific vision. Such accounts attempt to offer theological reflection upon last things without unduly constructing detailed renditions of either future events or overly confident descriptions of what features comprise eschatological life. While offering the necessary restraint in speculative matters, such accounts have more exclusively focused on the last things rather than the import of eschatology for moral praxis in the present. Whether the descriptions focus more on the exclusively human dimensions of eschatology[28] or attempt to include the broader realm of creation,[29] recent eschatologies have neglected the specifically moral relevance of eschatology for theology. In doing so, such authors have neglected the communal and relational connections present between the eschatological and the liturgical.[30] At times these eschatologies utilize language that denigrates current creation and minimizes its capacity to embrace divinity's manifestation. The living world as it exists is construed as the "devastation" rather than revelatory and replete with the capacity for eschatological presence.[31]

28. Thiel, *Icons of Hope*.

29. Griffiths, *Decreation*.

30. Ciraulo, "Sacramentally Regulated Eschatology in Hans Urs von Balthasar and Pope Benedict XVI."

31. Griffiths, *Decreation*, 4.

In contrast to this Buber argues that, unlike human history, nature receives God's creative act uninterrupted. "Nature, as a whole and in all its elements, enunciates something that may be regarded as a self-communication of God to all those ready to receive it."[32] However, in nature there too is violence, aberration, and violations of communities alongside harmful genetic mutations. Additionally, although humans arise from natural evolution and are a part of this world, the environmental destruction of this world has come at the hands of human seclusion from nature and a refusal of relational connectedness with the world. The alteration of canine genes through human breeding is one example of the alteration of the natural by humans. However, redemptive transformation also breaks in to those beings and changing relationships. The eschatological incursion of the between-space initiated in Bones's evangelization of me into life brings divinity near, but also highlights the tension found between eschatological presence and an unfulfilled present. This tension has been an essential theological tenet in the eschatology of Vatican II and in Protestant eschatologies that reflect intentionally upon the relationship between history, ethics, and politics.[33]

Eschatological life manifests the reconciliation of all things in God. In the life of a Christian, this reconciliation becomes a lived reality through peaceful relations, sacramental practice, and the ethical life exemplary of the kingdom of God expressed in the life and teachings of Jesus of Nazareth. Buber does not think that other animals have the ability to loosen the Gordian knot of our complex and broken existence. He denies the capacity of other animals to elevate a companion above the contingencies of their circumstance and forge something new.[34] However, like Buber's horse, my encounter with a small puppy attends to the manner in which eschatological life can interrupt the fragile and anxious ways in which we live. Such encounters remove the "delusional belief that for life we need only ourselves . . . that the realms of nature [are to] be bound and enslaved, made to do our bidding and satisfy our every wish."[35] Like Buber's horse, Bones is the instigator of relationship. But in this instance, he is more than Buber's horse. Bones is not simply entering into a relationship; he is reconciling a relationship that has been fractured.

Bones's eschatological act not only reconciled him to me and I to him. It also broke into the tension that he had experienced with humans.

32. Buber, *On Judaism*, 221.

33. Avliar, "La renovación de la escatología en el Concilio Vaticano II"; Moltmann, *Theology of Hope*, 103–4. See also Harvie, *Jürgen Moltmann's Ethics of Hope*, 13–37.

34. Buber, *The Martin Buber Reader*, 210.

35. Wirzba, "The Touch of Humility," 227.

He perceives humans differently and is no longer afraid of children of a certain age. It also sanctified my life in the world wherein I can no longer simply view the natural world as instruments of my use. Rather, the world is the beloved creation that is constantly being sustained and is struggling for its own amelioration. The daily liturgy of our shared practice continues to transform us and deepen our lives. "We do not touch each other as spectators. Our relationships with others, whether we admit it or not, are more intimate and involved than that."[36] Bones has grown into his vocation as apostle of grace through the vulnerability and courage of initiating the liturgical practices that allowed for reconciliation and the presence of the Other into our anxious lives. If the task of the theologian is to humbly and patiently think through the divine mysteries, then the reconciling, relational work of theology must be at the forefront of our endeavors with the hope that all things are renewed and made whole in the loving presence of God.

Bibliography

Abram, David. *The Spell of the Sensuous*. New York: Vintage, 1997.

Aristotle. *De Anima*. Edited by Richard McKeon. New York: Random House, 1947.

Atterton, Peter. "Face-to-Face with the Other Animal?" In *Levinas & Buber: Dialogue & Difference*, edited by Peter Atterton, Matthew Calarco, and Maurice Friedman, 262–81. Pittsburgh: Dusquesne University Press, 2004.

Aquinas, Thomas. *Aristotle's De Anima with the Commentary of St. Thomas Aquinas*. Translated by Kenelm Foster and Silvester Humphries. London: Routledge, 1959.

Avliar, J. José. "La renovación de la escatología en el Concilio Vaticano II." *Scripta Theologica* 46 (2014) 662–65.

Bremer, Józef. "Aristotle On Touch." *Forum Philosophicum: International Journal for Philosophy* 16.1 (2011) 73–87.

Buber, Martin. *Between Man and Man*. Translated by Ronald Gregor-Smith. New York: Routledge Classics, 2002.

———. *I and Thou*. Translated by Walter Kaufmann. New York: Simon & Schuster, 1996.

———. *On Judaism*. Edited by Nahum N. Glatzer. New York: Schocken, 1995.

———. *The Martin Buber Reader: Essential Writings*. Edited by Asher D. Biemann. New York: Palgrave MacMillan, 2002.

Ciraulo, Jonathan Martin. "Sacramentally Regulated Eschatology in Hans Urs von Balthasar and Pope Benedict XVI." *Pro Ecclesia* 24.2 (2015) 216–34.

Coppinger, Raymond, and Lorna Coppinger. *Dogs: A New Understanding of Canine Origin, Behavior, and Evolution*. Chicago: University of Chicago Press, 2001.

———. *What is a Dog?* Chicago: University of Chicago Press, 2016.

Harvie, Timothy. *Jürgen Moltmann's Ethics of Hope: Eschatological Possibilities for Moral Action*. Farnham, UK: Ashgate, 2009.

36. Wirzba, "The Touch of Humility," 237.

Hobgood-Oster, Laura. *A Dog's History of the World: Canines and the Domestication of Humans*. Waco, TX: Baylor University Press, 2014.

Garrett, Frank. "Extreme Humanism: Heidegger, Buber, and the Threshold of Language." *Between the Species* 10 (2010) 73–83.

Gita, Raimond. *The Philosopher's Dog*. New York: Random House, 2009.

Griffiths, Paul J. *Decreation: The Last Things of All Creatures*. Waco, TX: Baylor University Press, 2014.

Moltmann, Jürgen. *Theology of Hope: On the Grounds and Implications of a Christian Eschatology*. Minneapolis: Fortress, 1993.

Monks of New Skete. *The Art of Raising a Puppy*. Little, Brown, and Company, 1991.

Schaefer, Donovan O. *Religious Affects: Animality, Evolution, and Power*. Durham, NC: Duke University Press, 2015.

Schlosser, Julia. "Tangible Affiliations: Photographic Representations of Touch Between Human and Animal Companions." In *Experiencing Animal Minds: An Anthology of Animal-Human Encounters*, edited by Julie A. Smith and Robert W. Mitchell, 30–50. New York: Columbia University Press, 2012.

Tallon, Andrew. "Affection and the Transcendental Dialogical Personalism of Buber and Levinas." In *Levinas & Buber: Dialogue & Difference*, edited by Peter Atterton, Matthew Calarco, and Maurice Friedman, 49–64. Pittsburgh: Dusquesne University Press, 2004.

Thiel, John. *Icons of Hope: The "Last Things" in Catholic Imagination*. Notre Dame, IN: University of Notre Dame Press, 2013.

Wirzba, Norman. "The Touch of Humility: An Invitation to Creatureliness." *Modern Theology* 24.2 (2008) 225–44.

Wyschogrod, Edith. "Empathy and Sympathy as Tactile Encounter." *The Journal of Medicine and Philosophy* 6.1 (1981) 25–44.

My Life with Morris

A Feminist Account of Friendship and Conversion in a Bicultural Context[1]

—GRACE Y. KAO—

My relationships with cats began when I was a kindergartner. For reasons unbeknownst to me, stray cats and dogs would frequent our house on a cul-de-sac in a middle-class suburb of Los Angeles: they would turn up on our driveway every other month or so barking or meowing, as if pleading for us to take them in. My parents would allow us to care for the vagrant for a day or two, meaning that my older brother and I were permitted to play with it, provide it with temporary shelter in our garage, and feed it scraps of food or whatever else we could surreptitiously ferret away from our kitchen. But these joys were always short-lived, as my mom would eventually call animal control to take our new playmate away. We were always sad to see them go, though I remember feeling especially dejected when having to part with those cats.

Our fortunes changed for the better one Sunday morning when I was seven or eight years old. A member of our church had brought a box of kittens to give away for free from her neighbor whose cat obviously had not been spayed. As before, my brother and I begged my parents to let us keep one, pledging that we would do all the work. We were soon joined by a gaggle of kids our age beseeching their parents with the same promises. While all the other parents categorically refused, to our great delight and surprise,

1. I thank all three editors of this volume, Trevor Bechtel, Matthew Eaton, and Timothy Harvie, for their helpful feedback on earlier drafts and also Jeongyun Hur for research assistance.

our mom said "okay." She then directed us to select the largest of the litter, reasoning that the biggest would also be the sturdiest, as her frame of reference had been the scrawny feral cats who kept death at bay by scavenging the streets of Taiwan. So that was the day in the early 1980s when our family became part of the roughly two-thirds majority of American households who share their lives with a pet.[2]

We named our kitten Morris, as he resembled a tinier version of the most famous cat we knew—the eponymous orange tabby mascot of the popular line of cat food, 9Lives. I suspect that much of what I came to love about Morris was what other "cat people" adore about their felines: the velvety feel of their fur, the gentle vibrations of their purring, the way they suddenly interrupt their own normally calm demeanor with spasms of friskiness, and so forth. Still, Morris was unique to me for setting in motion my conversion years later to a life for animal others—a sketch to which I'll return in closing. What I'll elaborate on below, however, is the way Morris more immediately served as a window into my parents' world and thus my parents themselves, as our adopting Morris facilitated my ability to even see my parents as distinct individuals. More specifically, like their immigrant Taiwanese Christian peers,[3] my parents had largely regarded the practice of pet-keeping as an extravagant luxury—as a drain on or misuse of the family's time and money, which is why we previously were not allowed to keep those strays. But unlike their counterparts and more like their new countrymen, they nevertheless were now permitting us to indulge ourselves in this way.[4]

2. According to 2015–2016 American Pet Products Association Survey, 65 percent of households in the US include a pet. This two-thirds figure has remained somewhat constant for decades, as approximately 67 percent of American households in 1987 lived with a dog or cat. See Humane Society of the US, "Pets by the Numbers," and Clancy and Rowan, "Companion Animal Demographics in the United States," 2.

3. Studies have shown that there is ethnic variation in the relationships with and attitudes toward companion animals and that Asian Americans are the racial-ethnic group least likely to live with pets. See Risley-Curtiss, Holley, and Wolf, "The Animal-Human Bond and Ethnic Diversity."

4. Pet-keeping is on the rise in Taiwan today but that was not the case when my parents emigrated in the early 1970s before the dramatic rise in the standard of living following the "Taiwanese economic miracle" of rapid industrialization and economic growth. Today, news stories abound about Taiwan's first female president's love of cats and about the numerous "cat cafés" (places where patrons consume refreshments amidst free-roaming cats) that dot its big cities.

Caught Between Two Cultures

We raised Morris partially the Taiwanese way, consistent with the mores of our ethnic church community, and partially the American way, or as I had understood the matter from growing up stateside. Morris was like the large majority of (feral or stray) cats in the Taiwan of my parents' generation who lived exclusively outdoors, but he was dissimilar to them—and more like his American pet counterparts—in no longer having to forage for food to survive. (We of course fed him 9Lives canned food in honor of his namesake.) He was also more traditionally Taiwanese than middle-class American in never having been taken to the vet, though I hasten to add that he did receive excellent medical care when needed from my surgeon dad. Most importantly, he was "unclean" in my parents' eyes and treated as such as most animals were by Taiwanese folk then, but doted on and physically embraced by me—their American-born daughter. While the affection I showed him was utterly conventional when judged by mainstream American standards—I would pet him, scritch under his chin, pick him up and cradle him in my arms while sometimes kissing the top of his head—it was baffling to my parents and their friends in light of their characteristic Taiwanese reserve and restraint.

As I grew in awareness of the ways in which Morris was caught between two cultures, I came to see myself as similarly bicultural—as betwixt and between two ways of life. I accordingly oscillated between following my parents' wishes without hesitation and challenging them in a handful of cases, including on matters pertaining to my cat. Whether fair or not, I felt more Taiwanese when I obeyed them, given the Confucian-Christian virtue of filial piety and biblical commandment to honor one's parents, and more Western when I questioned or resisted them, in light of the stress on individual autonomy and ample examples of filial defiance I had seen among my (mostly white) classmates at school and in popular culture. The trick for me, of course, was figuring out when to comply and when to dissent.

For instance, I did not like how we never took Morris to the vet and worried at times that something unbeknownst to us might have been wrong with him. But I did not argue with my parents about it because doing so would have been futile. How did I know? They never even took my brother and me to the doctor for well-visits, despite our repeated requests to be taken (*n.b.*, we had gleaned from our white friends that that is what we were supposed to do), for my dad had analogously reasoned that he himself was competent to assess our medical needs and thus there was no need to go barring an emergency. Every few weeks or so, however, I would gently push back on something more important to me—their ban on permitting Morris indoors. Whether due to their own change of heart or my incessant pleading or both, my parents eventually caved and allowed Morris to come inside during the day if I took responsibility for him. From that point forward until I left for college, Morris kept me company near-daily on my lap as I read, did my homework, or even practiced piano.

Who Was Morris to Me?

It is empirically uncontroversial today that we humans can form emotionally significant bonds with some of the other animals with whom we come into regular contact. According to recent statistics in the US, pets even exceed the total human population: there are some 470 million pets scattered among 316 million people.[5] While owners or guardians still task their pets to perform a variety of functions including work, the ideal particularly in the case of cats or dogs remains for them to provide joy and emotional fulfillment through companionship. Many people concomitantly care for their pets with an intensity equal—or in some cases, exceeding—their care for their human loved ones, thus underscoring the familiar refrain that pets have become for them "part of the family."

While as a child I understood the intimacy and endearment that the notion of pets as family connoted, our Taiwanese heritage made it impossible for us to affirm this mainstream American understanding of animal companions as kin. This is not just because my parents left Taiwan in the early 1970s well before the Taiwanese began to conceptualize pets as appropriate outlets for emotional support (in the 1990s) or were even introduced to the concept of animals as pets (in the 1980s) with the rise of global market capitalism.[6] It is also because we as a family had internalized the importance of the "rectification of names" (正名, *zhèngmíng*)—a Confucian doctrine which teaches that right order and social harmony depend upon us calling things by their proper names, treating them accordingly, and behaving ourselves in ways appropriate to whatever title or office we held (viz., a mother acting motherly or a president acting presidential). Furthermore, family obligations in a Confucian context were prescribed—neither voluntarily undertaken, nor discretionarily relinquished—in role-specific ways. It would thus have been both conceptually odd and morally problematic for us to regard our cat like kin in any sincere way. That did not mean that we did not love Morris or anyone else not related to us, only that love alone

5. Pierce, *Run, Spot, Run*, 3.
6. Chang, "Trans-species Care," 289–90.

was insufficient to break the relatively nonporous boundaries between those who were genuinely family (and thus ought to be treated as such) and those who were not.

The fact that pets as kin has never worked for us for cultural reasons has led me to question whether another affiliative term could better capture the nature of my relationship with Morris. This I propose to do below through the category of friendship.

Friendship with Companion Animals

The appropriateness of calling Morris my friend obviously turns on what we take friendship essentially to mean or to be. While philosophers and theologians in the Western tradition have frequently turned to standard-bearer accounts for inspiration about its constitutive elements, this is not the path I take here given my feminist desire to draw upon, when possible, conceptual paradigms that foreground women's experiences. Mary Hunt's theological model of friendship not only does precisely this, but it also already incorporates nonhuman animals in its examples. As a result, conceptualizing nonhuman animals as friends neither requires an awkward application of her model, nor a direct countering of her stated position on the matter—things that cannot be said for those working instead out of most classical models (e.g., an Aristotelian or Thomistic framework).

In what follows, I apply and espouse a version of Hunt's feminist theological model of friendship. Its descriptive categories of analysis helpfully answer the question, "what *is* friendship?," and its normative categories of judgment help to answer an important follow-up question for ethics: "is the friendship *good*?" I ultimately conclude that conceptualizing other animals as (real or potential) friends not only enriches our understanding of the social bonds that are possible across the species, but also compels us to rethink the nature of friendship itself, particularly when significant differences or asymmetries between friends affect the kinds of relationships that can be developed or sustained.

The Model

Friendship on Hunt's score is an interaction of four components—love, embodiment, power, and spirituality. A friendship is said to work well when these elements are "present and in harmony"; it suffers when such elements are out of balance.[7] Though Hunt acknowledges that any model of friendship can be used reductively in ways that wrongly "mechanize[s] what is finally a mysterious experience," she offers this bare outline of friendship in the hope that its use will outweighs its risks.[8] As someone who has found her model helpful, I will elaborate below on its descriptive features while reflecting further on my life with my childhood cat.

First, I *loved* Morris, just as many others love or have loved their pets. Deep love for an individual animal even has scriptural warrants: there is nothing in the prophet Samuel's allegory (to David) about a poor man who loved his ewe lamb like a daughter to suggest that the man's tender loving care of his lamb was excessive or misplaced.[9] I myself loved Morris in some analogous ways as I loved my other (human) friends—I sought out his company, shared what I had with him, missed him when we were apart for long periods of time, and cared for him in ways I could when he was hurt or unwell. I also loved him not in spite of our differences (i.e., I was not hoping that he would function as "another self" as per Aristotle's model of the highest form of friendship),[10] but in many respects because of them. I found his otherness—his "catness"—charming, including the ways he would marvel at butterflies, meticulously groom himself from head to tail, and lounge around in the sunniest or warmest spots, be it on top of our water heater or our VCR.

While the notion that Morris loved me in return may be more difficult to substantiate given concerns about anthropomorphism, I have reason to believe in his reciprocation.[11] He would purr in my presence, therein showing contentment.[12] He would roll over and expose his belly to me, seek out my lap for naps, and look at me with half-closed eyes while slowly blinking—all behaviors that convey trust and enjoyment of my company. He would also occasionally leave dead birds on our doorstep. While as a child I wrongly interpreted those gory gifts as his way of gratefully paying us back for the food we provided him, I now know that these presents reveal that he not only saw us as bad hunters (i.e., incapable of catching such fare on our own), but also as part of his family constellation, since mother cats en-

7. Hunt, *Fierce Tenderness*, 99. Unless otherwise stated, all references to Hunt will be drawn from this text.

8. Ibid.

9. 2 Sam 12:1–10. For an intriguing defense of the appropriateness and even biblical mandate to love nonhuman animals, see Miller, *Animal Ethics and Theology*.

10. Aristotle, *Nicomachean Ethics*, 1166a.

11. For a defense of the existence of complex emotions in other animals, see Bekoff, *The Emotional Lives of Animals*; and Jonathan Balcombe, *Second Nature*.

12. For more on feline behavior, see Bradshaw, *Cat Sense*.

gage in similar behaviors when teaching their kittens how to hunt. Surely all these gestures were indicative or at least suggestive of love.

Secondly, just as most good relationships between humans and their pet cats or dogs involve forms of physical contact that are pleasurable to both parties, my friendship with Morris was definitely *embodied.* I spent a lot of time gently stroking his fur, manually defleaing him, and walking gingerly to avoid stepping on him when he would weave between my legs as I moved about the house. In turn, Morris would bunt (i.e., head-butt) me, rub his cheeks on me, knead me with this paws, or even groom me with his tongue—all acts that cat behaviorists say demonstrate affection and even ownership as felines mark their humans with their pheromones and sweat glands. My bond with Morris grew even closer (at least on my end) the few times I snuck him upstairs into my room at night to sleep in noncompliance of my parents' instructions. Just as touch and nonverbal communication take on "shared meaning with intimates" on Hunt's score, so biblical scholar Marti J. Steussy has called our primate-induced "deep yearning for physical social contact" a "sensual aspect of spirituality." In her words, "if stroking a cat or softly ruffling a bird's down elicits a happiness programmed deep in our genes, may we praise God for incarnation's joy."[13]

Thirdly, though domestication complicates the matter, both Morris and I exchanged *power* in our friendship, meaning that we each could make choices for ourselves. As human owners, to be sure, we had initiated the relationship simply by adopting him, and we continued to make unilateral decisions about major aspects of his care. Despite the asymmetry of power, Morris arguably retained a measure of freedom and even "negotiated" certain terms of our relationship. As an outside cat, for one, he could explore the neighborhood on his own, otherwise come and go as he pleased, and thus discretionarily refrain from having to be in our company. While we decided what food to feed him, he essentially trained us to dispense it at specific times through his persistent meows and taps at our patio door. He also demanded that I (and others) respect his bodily integrity: he would hiss, bite, jump, swipe, or scratch if he did not like the way he was being petted, bathed, or otherwise handled. Frankly, as a young girl who was being socialized into being ever-accommodating and deferential to others, I could have stood to learn from Morris's clear setting of boundaries, communication of wants, and unmistakable expression of displeasure when he didn't like the way he was being treated.

Finally, there were notable *spiritual* dimensions of our friendship. While all friendships for Hunt are theological for "illuminat[ing] questions

13. Steussy, "The Ethics of Pet-Keeping," 180.

of ultimate meaning and value," and for bringing us "face to face with ourselves, with one another, and with a larger world in which the mysterious forces of attraction work beyond our ultimate control," friendships with other animals additionally "reveal something about the oneness of the cosmos."[14] Namely, we learn from our nonhuman friends that the differences between and among us are not "categorical," thus challenging patriarchal, ontological, or otherwise hierarchical views that would prioritize humans on one side and all other animals on the other.[15] Indeed, Morris facilitated spiritual growth in me by "disrupt[ing] self-centeredness" and anthropocentrism and accordingly "inspir[ing] affection and appreciation for something [sic.] completely 'other'"—a fellow creature of the same God.[16] Beyond providing me with experiential knowledge of the goodness of all creation, there is another spiritual facet of my friendship with Morris. In mediating the love between my parents and me and in helping me come to terms with my bicultural identity, Morris was what theologian Stephen Webb has described as a "gift"—a clear example of divine grace at a time when I most needed it.[17]

An Assessment

To characterize a relationship as a friendship is not yet to provide a normative judgment about its merits or lack thereof. Hunt proposes that we prescriptively evaluate friendships on the basis of the four following fruits: attention, community, generativity, and justice. Hunt finds *attention* more important than trust, attraction, or admiration—it is that which "holds friends together" and allows friends to "slog on" in the midst of conflict or notable differences.[18] *Generativity* for Hunt means that which is new, brought forth, or develops beyond the friends themselves and thus would

14. Hunt, *Fierce Tenderness*, 83.

15. Ibid. As Hunt notes elsewhere, the differences between the "oddest combinations" of friends in a feminist construction "are not transcended but attended to with care so that the power differences are handled in a just way." Quoted in her entry on "Friendship" in Isherwood and McEwan, eds., *An A to Z of Feminist Theology*, 76. While I find much to commend in Hunt's theological description of human-animal friendships, my more conventional understanding of "spirituality" differs from hers, by which she means the process of "making choices about the quality of life for oneself and one's community" such that the "religious impulse toward meaning and value is expressed in very concrete ways." Hunt, *Fierce Tenderness*, 105.

16. Gilmour, "Companion Animals and Spirituality."

17. See Webb, *On God and Dogs*.

18. Hunt, *Fierce Tenderness*, 151.

not exist without the friendship's creative energies.[19] *Community* can form slowly and organically when attentive friends attract others who want to be drawn into their relational orbit or when persons with similar beliefs, commitments, or goals are motivated to care for one another. By *justice*, Hunt would have us assess both the treatment of the friends themselves as well as the effects on and potentially far-reaching implications for those on the outside.

What I like about Hunt's focus on *attentiveness* is that the ingredient that allows for friendships to flourish is the same virtue needed for character development more generally. As alluded to earlier, I had to observe Morris's behavior carefully (i.e., notice sudden changes in his eyes, ears, tail, or overall posture) to decipher when he wanted to interact with me and when he preferred to be left alone. Through feedback from Morris, I learned not only to discern what Morris as a cat needed or wanted—for example, opportunities to scratch and to climb—but also what Morris as an individual liked or disliked—for example, the 9Lives "super supper" variety that he apparently detested and simply would not eat. Marti Steussy has noted that most pet animals respond best to those humans who are "patient, self-controlled, fair, gentle, consistent, kind and able to understand that others, including these other creatures, may see things from a different point of view." She adds that this list "overlaps strikingly with New Testament descriptions of the 'fruits of the spirit.'"[20] Something analogous arguably applies to (same-species) friendships that cross gender, racial, religious, or other notable lines: a good friend genuinely attends to the other both as a member of a particular social group and as a unique individual who defies easy stereotypes. This is to say that the need for attentiveness—particularly to areas of difference—holds whether we are talking about befriending other humans or other animals.

The notion that cultivating a friendship with a particular animal can make one a better person is of course part of the appeal of petkeeping in families with young children. It also explains why the use of horses, dogs, cats, and other animals in animal-assisted therapy or rehabilitation (e.g., in prisons or hospitals) for both children and adults is on the rise: nonhuman animals can transmute stress, brighten spirits, and facilitate moral development in character as individuals caringly take responsibility for animal

19. Children begotten in love from a heterosexual marriage has been the classic example, though Hunt instead draws from women's friendships with other women to name other generative possibilities: protective spaces to survive (both psychically and physical), music, poetry, home repairs, theology, or whatever else that might emerge when "truths and talents that might have been left hidden" are given space to surface. Ibid., 151–62.

20. Steussy, "The Ethics of Pet-Keeping," 180. Emphasis added.

others. While attentiveness to Morris facilitated my growth as a human being in these ways, my friendship with him allowed me view myself as a particular kind of sociocultural being—a Taiwanese American. For as mentioned previously, Morris frequently served as the occasion for my explicitly learning Taiwanese ways; he likewise provided my family with opportunities to reflect upon how and why mainstream American attitudes—about animals, families, homes, preventative care, public displays of affection, and so forth—were different.[21] These interactions accordingly allowed me to appreciate the wisdom and the excesses of both cultures and therein come to affirm my bicultural identity.

That we became full-time pet-keepers at all and even came to share the interior of our house with Morris are some of the fondest memories I have from childhood, in part because they were concrete ways that my parents turned toward my brother and me in love.[22] These moments also functioned as mini-epiphanies, for I learned not only that my parents could change their mind even on matters about which they were culturally conditioned to feel strongly, but also that I had some capacity to persuade them. Indeed, my parents came to accept my influence in other matters pertaining to Morris, including by eventually emulating me in cuddling him affectionately in ways they would previously never have done. In short, my friendship with Morris prompted my parents to develop their own caring relationships with him, therein generating a greater sense of intimacy, love, and *community* among us all.

In addressing the *justice* dimension of my friendship with Morris, I return now to the conversion to a life for animal others that I referenced earlier. According to Hunt, good friends of animals work to "extend the demands of justice to all creation" in various ways, such as by safeguarding certain species, eating lower on the food chain, campaigning against "cruel hunting, vivisection, and unnecessary animal experimentation," and

21. In so doing, Morris played what bioethicist Jessica Pierce has called the "social lubricant" role in providing "bridges or links to other potential friends and help[ing] people initiate social interactions." See her *Run, Spot, Run*, 62.

22. For instance, though I was overjoyed then when my parents eventually relaxed their no-animals-inside policy, I only came to understand years later that their concession—a counter-cultural violation of taboo, if you will—was a real gift of love to me. For like most East Asian families, we followed a strict no shoe policy in our home: all shoes and slippers were to be worn outdoors unless they were specifically designed and designated for at-home wear. Thus, to allow Morris to have free egress inside and outside of our home, without having to undergo any sort of cleanliness ritual (i.e., of removing shoes, so as not to track dirt inside), was to privilege him in ways we did not even treat our guests or ourselves.

seriously considering the rights of nonhuman animals to "life and limb" along with human rights to the same.[23] So understood, while I had befriended Morris in my childhood (and he, me), I did not become a good friend of animals until decades later. Morris was nevertheless instrumental in putting me on a long path of greater concern and advocacy for animal others. As in the case of other scholar-activists I know, my slow conversion toward animal others was ultimately a matter of "head" and "heart": my scholarly interest in nonhuman animals emerged organically out of my study of human rights and ecofeminism; the love and fond memories I had of Morris and later Puzzle (the calico cat I adopted as an adult from a no-kill animal shelter) subsequently "added the power of feeling to the requirements of logic."[24] When I teach, publish, or render service in groups in my scholarly societies that study and advocate for nonhuman animals, I do so now with empathy and passion—things I first experienced with Morris.

Conclusion and Postscript

In conceptualizing Morris as my friend and then describing how and why the friendship was good, I have shown in ecofeminist fashion how a model originally constructed from a feminist theological reflection on women's friendships but also originally extended to cover other animals applies in my case. Love, embodiment, power, spirituality, attention, generativity, community, and justice—these are all important dimensions in friendships. Friendships with nonhuman animals not only disclose exciting possibilities for interspecies relationships, but they also prompt further exploration of the constitutive requirements of friendship itself. More specifically, the "obvious" questions that human-animal friendships raise (e.g., do domesticated animals have any real power? To what extent are these relationships voluntary? Are humans simply projecting emotions onto their animal friends?) should propel us to ask non-obvious questions about friendships among humans themselves (e.g., how important theologically is it for friends to be genuine peers, given that Jesus called his disciples "friends"?).[25] Do we really "choose" our friends or are there other forces—such as attraction, luck, and coincidence—at work beyond our immediate control? Is not the concern about false attribution and projection just as real in the human case as it is with other animals?

23. Hunt, *Fierce Tenderness*, 174, 83.

24. Regan, "The Bird in a Cage," 93.

25. John 15:15.

While the "justice" potential exists in our friendships with particular animals, we should nonetheless remain modest about where these friendships will actually lead. For not only does the concept of friendship generate only conditional, not categorical obligations (i.e., we cannot obligate someone to befriend anyone else), but we humans have also historically shown great inconsistency in our treatment of other animals—a fact made clear by title of a recent book on the psychology of human-animal relations: *Some We Love, Some We Hate, Some We Eat.*[26] Just as a genuine friendship between an Asian American and an African American won't necessarily compel the former toward social justice issues involving the latter's community, so it remains to be seen whether friendships with particular animals will lead a person to advocate for justice for them all. Still, we can hope that a satisfying and meaningful relationship with an animal friend can give one an abiding emotional, not just rational, motivation to expand one's circle of concern in ever widening ways, as it did in my case.

Bibliography

Balcombe, Jonathan. *Second Nature: The Inner Lives of Animals*. New York: St. Martin's, 2010.

Bekoff, Mark. *The Emotional Lives of Animals*. Novato, CA: New World Library, 2008.

Bradshaw, John. *Cat Sense: How the New Feline Science Can Make You a Better Friend to Your Pet*. New York: Basic, 2013.

Chang, Chia-Ju. "Trans-species Care: Taiwan's Feral Dogs and Dog Mother Activism." *International Journal of Humanities and Social Science* 2.3 (February 2012) 287–94.

Clancy, Elizabeth A., and Andrew N. Rowan. "Companion Animal Demographics in the United States: A Historical Perspective." In *The State of the Animals II: 2003*, edited by Deborah J. Salem and Andrew N. Rowan, 9–26. Washington, DC: Humane Society, 2003.

Frööding, Barbro, and Martin Peterson. "Animal Ethics Based on Friendship." *Journal of Animal Ethics* 1.1 (2011) 58–69.

Gilmour, Michael. "Companion Animals and Spirituality." *Huffington Post*, November 15, 2011. Accessed April 17, 2017. http://www.huffingtonpost.com/michael-gilmour/companion-animals_b_1083096.html.

26. Philosopher Mark Rowlands essentially raises this critique against two philosophers who have not only attempted to extend Aristotle's views on friendship to the case of other animals, but also argued for the moral significance of having done so. For this exchange, see Frööding and Peterson, "Animal Ethics Based on Friendship," and Rowlands, "Friendship and Animals." Psychologist and co-founder of the field of anthrozoology Hal Herzog published the aforementioned book about our irrational and paradoxical treatment of nonhuman animals with Harper Collins in 2010.

Herzog, Hal. *Some We Love, Some We Hate, Some We Eat: Why It's So Hard to Think Straight About Animals*. New York: Harper, 2010.

Humane Society of the US. Last accessed November 15, 2016. "Pets by the Numbers." http://www.humanesociety.org/issues/pet_overpopulation/facts/pet_ownership_statistics.html.

Hunt, Mary E. *Fierce Tenderness: A Feminist Theology of Friendship*. New York: Crossroad, 1991.

———. "Friendship." In *An A to Z of Feminist Theology*., edited by Lisa Isherwood and Dorothea McEwan, 74–76. New York: Bloomsbury Academic, 2016.

Miller, Daniel. *Animal Ethics and Theology: The Lens of the Good Samaritan*. New York: Routledge, 2011.

Pierce, Jessica. *Run, Spot, Run: The Ethics of Pet-keeping*. Chicago: University of Chicago Press, 2016.

Regan, Tom. "The Bird in a Cage: A Glimpse of My Life." *Between the Species* 2.2 (1986) 42–49, 90–100.

Risley-Curtiss, Christina, Lynn C. Holley, and Shapard Wolf. "The Animal-Human Bond and Ethnic Diversity." *Social Work* 51.3 (2006) 257–68.

Rowlands, Mark. "Friendship and Animals: A Reply to Fröӧding and Peterson." *Journal of Animal Ethics* 1.1 (2011) 70–79.

Steussy, Marti J. "The Ethics of Pet-Keeping: Meditation on a Little Green Bird." *Encounter* 59.1–2 (1998) 177–95.

Webb, Stephen. *On God and Dogs: A Christian Theology of Compassion for Animals*. New York: Oxford University Press, 1998.

FARM AND LAB

Born to Be Wild?

Emergent Wisdom through Human-Horse Encounters[1]

—CELIA DEANE-DRUMMOND—

In keeping with the intention of this volume, I am going to begin this chapter with personal experiences that I have had with ponies and horses, concentrating on the formative years between the ages of eight and eighteen, when I was fortunate enough to be in a family that was committed to owning and caring for ponies, and eventually horses, three dogs, and a cat.[2] My mother even started breeding horses once my sister and I left home; Connemara ponies crossed with thoroughbred horses, leading to strong and desirable riding horses that had both stamina and elegance. My sister and I were both given ponies at an early age in order to improve our equestrian skills after we had learned basic skills in a riding school. My mother also owned her own pony, called Penny Bright, and joined with us in exercising and other activities, though she did not take part in any of the competitions that we regularly entered.

My sister, Anna, was particularly talented in horsemanship and regularly won cups, prizes, and other awards both in basic show jumping, but

1. I would like to thank anthropologist Dr. Marcus Baynes Rock for helpful feedback on this chapter, as well as his generosity in referring me to published works on this theme, and for supplying crucial ethnographic information that was extremely helpful in composing this chapter. I would also like to thank Lorraine Cuddeback for superb editorial assistance and the editors of this book, Trevor Bechtel, Timothy Harvie, and Matthew Eaton, for inviting me to contribute to this volume.

2. In keeping with equestrian jargon I am using the term *pony* to refer to a horse that is below 14.2 inch hands (measured at the horse wither from the ground up). I only owned a horse at the very end of my horse ownership, between ages sixteen and eighteen.

also the more advanced sports such as dressage and hunter trials. Quite simply, life outside our school education was totally dominated by our relationships with our ponies.[3]

The first pony that I remember clearly owning at around age eight was called Benjamin, a New Forest dappled bay gelding that was very safe to ride. The bonds of affection were strong, and I considered Benjamin to be one of my best friends. He was very calm and good-tempered, and always came up to me, nuzzling me affectionately when I went into his field. He knew very well that he was my pony, and I took him to events called gymkhanas where we entered basic schooling skills competitions, such as walk the poles, best kept pony, and so on. There was no doubt at all in my mind as a child that Benjamin had a personality, knew me by name, communicated with me both through his bodily movements and sometimes vocally by a whinny, and was intelligent.

Eventually I became too tall to ride him, and as my equestrian skills advanced I needed a pony that was more challenging to ride. Kelpie filled this role when I was about ten years old: he was a light bay gelding with a white stripe down his face. Kelpie was livelier than Benjamin; he was still a relatively easy ride, but was also willing to jump fences. In learning to ride I knew exactly what kind of mood Kelpie was in when we entered the ring in a competition, and whether or not we were likely to come away with any prizes. Kelpie sometimes used to do what is called "refusals," that is, come up to a fence and then stop suddenly. It would be very easy for me to fall into the fence, but I learned to keep my balance. I found that it was often a battle of wills: my will against Kelpie's, and that was communicated through my body's movements in the saddle. My sister, who was a more experienced rider than I was, and also very talented, could almost always get Kelpie to do things that I could not.

When I was around twelve or thirteen years old Kelpie became too small for me, and we acquired Diamond, who was a New Forest dark bay gelding with a clear star on his face, another smaller one on his nose, and a white "sock" on one of his back legs. I adored Diamond, and I immediately knew his mood from the way he positioned his ears, his affectionate nuzzle

3. Anna's passion has always been for horses, and she has kept sequential horses after adulthood. As a young woman she considered becoming a professional rider as a career. When she married, her husband learned how to ride. She currently lives in Devon in the UK, where she still keeps horses and dogs. Her daughter graduated from university with a degree in equine science. The childhood impact of being in the horse world that I shared with her has never left Anna, and has had a formative influence on her life and decisions, as well as those of her daughter, Rebecca, who is currently training to work as a professional horse physical therapist. This horse world forms a close-knit community for those inducted into it.

to me, and his willingness to jump higher and yet higher fences. He took me to heights I never thought would be possible for me, and gave me confidence that we could make it together as a team. Our efforts arose from a close and intense relationship, and I never had any doubt that Diamond was a partner with me in every event we entered together.

When I was sixteen years old, due to my growing height, I had to move to ownership of a horse, and, for that I acquired a 15.2 hand Irish light-bay mare called Shamrock. She had a white blaze, like Kelpie, and four white socks. Shamrock was everything you might want in a horse: solid, dependable, reliable, willing to jump, and affectionate. Horses are not quite as intimate with their owners as ponies, on the whole, and in this respect my relationship with Shamrock was not as intense as it had been with Diamond. I did, however, take meticulous care in getting her ready for shows; plaiting her mane, washing her tail, grooming her for hours until her coat shone brightly. I loved her, and my own sense was that the feeling was mutual. It broke my heart that my mother had to sell her when I went to Cambridge University for my degree in natural science. Nonetheless, all these encounters helped to shape me as a human being: my priorities, my affections, my interests. Although I did not make riding a career aspiration as my sister decided to do, it was still deeply formative in that my intimate encounters with ponies and horses helped to shape my sense of values in my youth through to my adolescent years.

Human-Horse Relations

To what extent are the experiences that I described above common? Are they musings and distant memories of a naïve child that are not worth much more than nostalgia? And what might this mean from the perspective of theology and ethics? Certainly, these horse encounters are deeply lodged in my memory, with as much vividness as any close or intimate human relationship. In fact, the constant contact over the years meant that these bonds with ponies and horses were even deeper, in many cases, than those I shared with my friends. In order to test the reflections I have outlined above, I am going to look at two sets of literature: one from social science and psychological studies of horses, and one that involves ethnographic studies with Oromo people. The latter is as yet unpublished, but it is gleaned from information provided by Marcus Baynes-Rock, an experienced anthropologist and ethnographer who is currently in our research team at the University of

Notre Dame, researching processes of domestication and the implications of this for the evolution of complex human social intelligence.

To begin with, there is recognition in the social science and psychological literature that horses and ponies have strong personalities, and that these do not simply arise from contact with humans or as a projection from our own anthropomorphic sensibilities, but rather are important even among con-specifics. Horses and ponies are highly social herbivores that in the wild, or where possible in domesticated settings, live in long-lasting groups. The research suggests that personality types of particular horses will influence the place of that horse in a collective group hierarchy, with positive affiliations forming between those horses or ponies that have a similar personality and place in the hierarchy.[4] This study was the first of its type to show positive assortment according to personality type in a non-kin group within a mammalian species.[5] Most of the collective movements are initiated by bolder individuals, rather than simply according to rank in the hierarchy.[6] This research coheres with my own memory of how our horses behaved in a much smaller pack of three or four in a smaller area; it was always the dominant individual who initiated movement to new pastures at different places in the paddock.

Another fascinating research article on the psychology of horses has shown that there is evidence that horses will respond to facial images in humans expressing happy or angry emotions.[7] Horses displayed left-gaze bias[8] (which, in the particular horses studied, is cognitively a right hemisphere lateralization that is especially sensitive to stimuli perceived to be negative), as well as a quicker rise in heart rate when presented with

4. The study was conducted by 200 hours of observation on a group of thirty-eight horses living on a thirty-acre hilly pasture. Briard, Dorn, and Petit, "Personality and Affinities Play a Key Role in the Organization of Collective Movements in a Group of Domesticated Horses."

5. This is asserted by the authors. Ibid.

6. Although it seems likely that collective patterns are influenced by both social determinants such as hierarchy and personality differences, the Briard et al. study did not show how these two factors were related.

7. Smith et al., "Functionally Relevant Responses to Human Facial Expressions of Emotions in the Domestic Horse (*Equus caballus*)."

8. Left gaze bias (LGB) is demonstrated in human subjects whose gaze is bias towards the left visual side of faces (right side of the face of the subject being viewed). Rhesus monkeys show a left gaze bias towards humans and monkeys, while dogs show a left gaze bias just towards humans. This bias develops in humans and is undifferentiated in six-month old babies. Facial communication is an essential part of social cognition. The right hemisphere of the brain receives images from the left-hand visual field, and this hemisphere is more sensitive to perceptual processing. See Guo et al., "Left Gaze Bias in Humans, Rhesus Monkeys and Domestic Dogs."

photographs of angry faces. These results are particularly interesting as it implies that there are genuine interspecies exchanges of information based on emotional states. A similar lateralization response to human angry faces has been found in dogs.[9] This study resonates with my own experience that ponies and horses picked up the mood of their riders almost instantly, and responded according to their own personality type. A nervous pony, for example, would generally be more sensitive to angry moods than a pony that was known to be calm and collected. Positive responses to happy moods were harder to determine, and the authors suggest that this may be because the horses were unfamiliar with the images presented. This would make sense, as detecting an angry mood in a stranger would be more beneficial in an evolutionary sense than a happy mood.

In a third study, Keri Brandt has used ethnographic data of interviews of those who are heavily invested in relationships with horses.[10] Her results cohere in broad outline with my own experiences, in so far as she stresses the way the communication system between people and horses develops through whole body movements and contact. Given the size of horses, and the fact that they cannot speak human language, other ways of communication are brought to bear, the most significant of which is sensitivity to touch. She argues, in particular, that symbolic interaction is possible without use of language. As her article suggests, in my own experience and that of her interviewees, horses are extremely sensitive to the position of the body in the saddle or on the horse; they "know" immediately if a rider is experienced or not, and will act or react accordingly. A calm horse will generally stay calm, but a nervous horse will play up, or even buck the rider out of the saddle. Missy, a woman Brandt interviewed, said "you can feel nervousness up through their back . . . . Through your body."[11] This has been my experience, as well. Naturally, the tendency is to then react with nervousness; a good rider will remain calm and communicate that mood to the horse. Furthermore, a calm horse can communicate that to a nervous rider, too. I have helped those with learning disabilities gain some equestrian experience, and the results show the positive, and even therapeutic, benefit of physical contact with horses.

Like Brandt, I also agree that the ears of a horse are very expressive of their moods. If you are riding a horse and it lays its ears back, and starts to buck, you are in real trouble. This is an angry horse. As Jane, in one of Brandt's interviews, describes it: "the ears will be different. You know, one

9. Racca et al., "Reading Faces."

10. Brandt, "A Language of Their Own," 299–316.

11. Ibid., 306.

good little squeal and a buck, that's exuberance. Ears back and rooting you out of the saddle and bucking. That's pissed off."[12] So, anger can be communicated to the rider from the horse, as well as the other way round. I do, however, question the extent to which Brandt seems to deny any possibility of verbal communication using language between rider and horse. A very well-trained horse will respond to verbal commands such as "canter," "trot," and "walk" by an observer on a ground with a lunge rein, with little or no physical cues; albeit the cues normally have to be associated with the tactile movement, first. It is also quite possible that each verbal command may be associated with very small changes in body language. For a human participant, however, the experience *feels like* the horse understands human language, even if it is picking up on other cues.[13]

Some individual horses are known to be more "moody" than others, and some breeds are habitually more nervous than others. Trainers regularly recognize the different character of horses and attribute such characteristics to them, and Vicki Hearne has demonstrated as much in her discussion of horse training, sensitive horses and crazy ones too.[14] This indicates that large differences in relationship are possible, depending on the particular horse or pony. Sometimes a very nervous and disobedient horse becomes like that through a history of abuse by its owners, analogous to personality disorders in humans after mistreatment. Ear movements will often pick up the mood of horses, and unlike the lateralization study of Smith et al., discussed earlier, ear behavior seems to be sensitive to positive moods. A behavioral study with forty veterinary students showed that a positive attitude and loose leash tension towards a specific horse was associated with infrequent ear movements and forward-facing ears in comparison to loose

12. Ibid., 308.

13. A famous example of how the sensitivity of the horse to body movements fooled the scientific community is that of "clever Hans," who stomped his hoof in response to specific questions about the time on clocks and other remarkable cognitive acts. Later, it was found that if questions were asked without the questioner being visible, his success collapsed. He was apparently sensitive to the small movement of the head of the person asking the question when they articulated the correct answer that Hans then identified. See Wynne, *Animal Cognition*, 9–12. Vicki Hearne, on the other hand, has challenged the extent to which intellectuals refuse to countenance the fact that in training verbal language is used regularly to communicate with particular animals in particular ways, so views explanations such as that of Wynne's with a degree of suspicion. She argues that the fact that the horses, dogs, and so on are sensitive to the morally loaded language of the trainers was itself significant, most particularly because the training works practically in ways that other methods do not. See Hearne, *Adam's Task*, especially introduction and the chapter on Crazy Horses.

14. Hearne, *Adam's Task*.

leash tension alone.[15] As this study was conducted on one horse, it is hard to have complete confidence in the results, but it fits with frequent anecdotal evidence from interviews with experienced trainers.

It cannot be overstated that in considering human-horse relationships, the horses should not simply be viewed as passively reacting to the moods of their owners. Ethnographic data shows up the difficulty of the categories of "wild" and "tame," especially as this is applied to thick ethnographic studies of horses, which transgress wild/tame, nature/culture, mutuality/control, free/not free binaries.[16] Furthermore, in the anthropological literature, the variant of communalism known as *biosociality* suggests that humans are viewed as an interface with and in nature, where practical acts and interactions in the wider communities trump abstract body/mind separation.[17] This need not dissolve all distinctions, but they are distinctions crafted in a new way, without the baggage associated with earlier dualisms.

Ethnographic research resonates with my experience, that close association with horses has a powerful and formative effect on the rider's sense of self. Anna, my sister, always claimed that she liked horses more than people, even though her own professional work in personnel relations and senior management for a supermarket involves highly developed people skills. She is not alone, and there are dozens of reports of similar attitudes. Dona Davis et al. interviewed a number of riders through a participant observation study. Xena, a dressage rider, claims, "Horses are who I am. They become such a part of you they make you who you are."[18] A more explicit example of moral development is illustrated from an interview with an endurance rider named Black Bear in the same study: "My daughter Kate is who she is today because of her relationship with horses. It made her a young woman who is not afraid of anything, is willing to take on any challenge, and simply refuses to be told something cannot be done."[19] The Norwegians in the sample were less revelatory but said something similar, as Ola comments "since childhood the horse has been stuck to my skull."[20] The key emotions in such

15. Chamove et al., "Horse Reactions to Human Attitudes and Behavior," 323–31.

16. Elizabeth Lawrence discussed such transgressions over thirty years ago in her book *Rodeo.*

17. Tim Ingold argues that biosocial is a radically new way of considering the biological and social: not so much as separated or related entities, but interwoven, so that "the domains of the social and the biological are one and the same." Ingold, "Prospect," 9.

18. Davis, Maurstad, and Cowles, "'Riding Up Forested Mountain Sides, In Wide Open Spaces, and Without Walls.'"

19. Ibid., 58.

20. Ibid.

human-horse relationships are those of empathy, connectedness, and trust. Event riding, which my sister took part in, involves a high degree of courage by rider and horse, as well as intense communication. Dona Davis points out that even within different sports there are subtle differences in how the human-horse relationships develop, so riding in an indoor space is a very different experience from riding outdoors; and dressage is very different from cross country events. These differences suggest the need to pay close attention to context.[21]

Oromo Horsemanship

Oromo is an ethnic group making up about 40 percent of the population of Ethiopia, and found more widely in the horn of Africa.[22] My specific interest in this section is to further draw out the subtlety of the relationships between horse and rider in a very different cultural context from those that I have discussed so far. Historically, in the Gibe region west of Shewa where Oromo kings took their name after their horses. Abba, which means father of, is followed by the horse's name. Abba Jifar means father of dappled; Abba Magal means father of dark bay; Abba Boqa means blaze on the forehead.[23] Anthropologist Marcus Baynes-Rock is married to Tigist Teressa, who comes from the Oromo group living in the Koro Hari region. In an interview with her about Oromo life she explains how the close relationship with their horses starts right from birth, with mothers riding horses with the baby on their backs.[24] There are two kinds of horses in Oromo culture; the *farrda mia*, who are used for racing or presiding at large-scale funerals, and the ordinary pack horses. One example of a funeral procession was captured on video following the death of Meles Zenawi, when a group

21. Ibid., 62.

22. There are several different subgroups within the ethnic group of Oromo, but all trace their origin back to the common ancestor, the mythical *Orma*. The Oromo and their horses still live in close affiliation in the volcanic and mountainous Koro Hari region of Ethiopia in a way that informs and shapes their lives in highly significant ways. The history of association between Oromo and their horses goes back to the sixteenth century, when their warlike raids on neighboring communities in a period of rapid expansion benefited considerably from the speed and mobility of horse companions. Baynes-Rock, personal communication, April 16th 2016.

23. Ibid.

24. Marcus Baynes-Rock, "Interview with T. Teressa on Oromo Horses." I am very grateful to Marcus and his wife for their willingness to share this information much of which is published in Baynes Rock and Teressa, "Shared Identity."

of Oromo men rode to Addis Abba and performed a traditional funeral ceremony.[25] The calm behavior of the horses amidst the noise and bustle of this event was remarkable. The Oromo bring their *farrda mia* horses, which are highly prized and considered part of the family, even bringing them into their houses for shelter. Sometimes the packhorses come in as well, due to fear of predators like hyenas or thieves, but "they [pack horses] can go crazy in the house but the *farrda mia* is alright."[26]

Between the ages of three and five, children ride bareback with a rope bridle. By nine or ten they are galloping bareback, holding onto the mane and swung all over the horse's body. It is as if the child becomes an extension of the horse's own body; such is the bodily intimacy. These horses are pack horses rather than *farrda mia*. Just as in the common experience of horse sensitivity to their riders noted earlier, the *farrda mia* horses do not take to having children on their backs; they know right away, and according to Teressa "they're not settled at all," and "think it is a bird." In this case, the horses are alienated from what is felt to be the "wrong" person. My own experience with horses, borne out by Brandt's research discussed earlier, is that horses know almost immediately upon taking the saddle or climbing on their back how experienced the rider is, and also recognize their own habitual rider. As Vicki Hearne states unambiguously "They *know* when you're afraid."[27] This also works the other way round, that is, how far a rider is prepared to trust a horse; as in famous examples of riders who gave free rein to horses who could sense where mines were placed, discussed by Hearne.[28] The *farrda mia* horses know their riders intimately, and in some cases will refuse not just to be ridden but also to be caught by anyone else. Teressa explains,

> Sometimes you tie it up to something and go and get something, someone comes and tries to take the horse, he jumps around. That one is a good horse, a special horse. Some horses go with anyone. I don't know how they know or how they train them. And that's one of my dad's favorite horses. Anyone trying to take him or trying to catch him, he'd go crazy, even my mum never caught him.[29]

25. The funeral procession of Shewa Oromo performed when Meles Zenawi died is shown in the following video: Ethiopian Television, "Ethiopians in traditional Oromo costume pay their respects to PM Meles Zenawi."

26. Baynes-Rock, "Interview."

27. Hearne, *Adam's Task*, 78.

28. Ibid., 24.

29. Baynes-Rock, "Interview."

The *farrda mia* horses are dressed up in costumes for racing or other occasions such as funerals, are trained in order to pace in a specific way, and are ridden by the menfolk who ride with shields and spears.[30] The racing itself is very dangerous and sometimes fatal, since spears are used to dislodge other riders from their saddles; though today poles have replaced spears, leading to lower fatalities. *Farrda mia* horses are given barley grain as feed, a delicacy that is even denied to the children. The men take great pride in combing and grooming the horses, and in preparing for the race by abstaining from any drink. The man in a horse-rider pair that wins a race becomes highly sought after by women as a potential partner. One of the great benefits of winning is the reputational prestige that this brings, and the admiration for those who win in these contests. There are different races depending on the ages of those taking part, so *Timkat* is for young people, while *Mezkel* is for mostly older, married people. An Oromo man is also ready to admit that he calls his horses each by name and in one case "loved him very much."[31] There was a mutuality, so the horse did whatever he desired, and seemed to know what that meant. In this way the human relationships of the Oromo, even the most intimate ones like marriage, are bound up in skills of horsemanship and human-horse communication.

Teressa is also explicit about the way ordinary human language is used to communicate with the horses, as well as physical cues. So, people talk to horses about "anything and everything" as part of building up the bond, and, in her experience, they understand what is being communicated verbally. So, "when you say, 'Let's go walk' then he jumps and gets excited and then comes. As soon as you come with your comb and say, 'Oh you want to be washed up' . . . Walk out and come." As to whether the horse understands the words or just the sound of the voice, Teressa admits, "I don't know."[32]

Teressa had a nasty accident with a horse when she was young, and was dragged along hanging by the stirrup for some miles, bouncing by one leg. Fortunately, it was on boggy ground, otherwise it could have been fatal. Given this horrific experience, it is not surprising that the favorite horse in her memory is her aunt's packhorse called Booqee. She explains:

30. An institution called *Gada* is the historical background. Within the *Gada* system there are eight grades, and men in all Oromo clans progress through these grades, moving to a higher grade every eight years. In each grade they hold new responsibilities and all those in one grade are considered equal. Historically in transition to a new eight-year cycle those who are of the age to practice warfare are obliged to attack a neighboring community outside the Oromo ethnic group called a *butta* war. A man's conduct in *butta* defined the status of that man for the rest of his life.

31. Baynes-Rock and Teressa, ethnographic field notes prepared for Bayes Rock and Teressa, "Shared Identity".

32. Baynes-Rock, "Interview."

> He's very high, big. He's gentle, he's not like horses [that] jump around. He's not a racing horse, he's more [for] carrying things. And he's huge and you talk to him. He knows all the family sound from a long way. You call him, he comes. You just go and call, "Booqee come home!" And he'd come. He chases you home and it's cute. And [he] play[s] with you. He'd chase you home and like chase you and then you stop and the he'd come.[33]

For Teressa, the horse she loved best was not one of the highly decorated and high flying *farrda mia* horses , but an ordinary pack horse; Booqee, a gentle giant who could communicate with her, and with whom she became friends. So, while love between Oromo men and their favorite racing *farrda mia* horses could be viewed as the result of the prestige and status for the men and their respective families, the love that Teressa showed was just for the sake of the relationship itself, rather than for any additional social benefits in the human community. The point here is that the dense interactions between humans and horses have a lasting impact on Oromo choices and behavior. Further, these relationships are complex, ranging from bonds arising out of familiar contact as a child, and (in the case of *farrda mia* bonds) cemented through the social prestige and the status that this relationship brings.

Towards An Interspecies Ethics[34]

In other publications I have started to trace out the possibility of what I have termed *inter-morality*, or interspecies morality, signaling the multispecies domain as a site for navigating complex human relationships.[35] In other words, human relationships do not exist alone, but with other creatures, even in sophisticated political societies such as those in the Western world, as Cary Wolfe has elegantly argued.[36] I have also argued that in an evolutionary sense the ability to navigate complex relationships is a form of *wisdom*, and that wisdom includes those other creatures with whom we have learned to live alongside through centuries of contact, including domestication.[37]

33. Ibid.

34. See, for example, Cynthia Willett, *Interspecies Ethics*.

35. Deane-Drummond, "Deep History, Amnesia and and Animal Ethics"; Deane-Drumond et al., "The Evolution of Morality"; Deane-Drummond, *The Wisdom of the Liminal*.

36. Wolfe, *Before the Law*.

37. Deane-Drummond, *Wisdom of the Liminal*.

Philosophical hygiene demands that we define terms and are careful with concepts; the term *morality* is very slippery and hard to pin down in this respect, since the way it is used by moral philosophers or theologians is very different from the way the term is used, for example, in evolutionary psychology, or even moral education.[38] Some philosophers have objected to the idea that animals other than humans have a sufficient theory of mind to be considered free agents, and therefore should not be thought of as having "morality" of any sort. According to this definition, morality requires a type of reasoning and detached freedom of choice that is thought only to be possible for human beings. Others, myself included, argue that in a generous reading at least *some* social animals seem to make deliberative choices according to certain patterns or rules set up in their own social spheres. Their rules are not the same as those for human societies, but nonetheless some boundaries are there. While this may not approach the sophisticated forms of justice found in human communities, at least nascent forms of what looks like wild justice are worth careful consideration.[39]

Further, the evolutionary emergence of wisdom understood as the competent ability to navigate complex sociocultural relationships is present in other animals in addition to humans, and the evidence for such navigation is borne out by frequent examples in the literature, not least the human-horse encounters I have described here.[40] That process of domestication goes as far back as the Holocene era, when humans learned to create and navigate highly complex social niches that included relationships with specific animals, both domesticated and predatory.[41] Baynes-Rock argues that human perception that other beings have minds like our own, that is, of "mindedness" in the nonhuman world, including that in other animals in particular, is both a compulsive and adaptive tendency in human communities. This observation avoids the quandary of whether or not and in what sense such a theory of mind in other animals is justified; the point is that human communities learned to perceive animals in this way, and that this perception was adaptive for survival. For example, if ancestral humans were scavengers and competing for food from carcasses with other predators, then (especially in the absence of language) the ability to work out what the other agent might be thinking would be a competitive advantage. Further, what is particularly interesting in the case of the domestication of

38. For a discussion of three different approaches to the evolution of morality see Deane-Drummond, Fuentes, and Arner, "The Evolution of Morality."

39. Deane-Drummond, "Natural Law Revisited,"

40. Baynes-Rock, "In the Minds of Others."

41. Ibid.

horses, goats, and sheep in the West Shewa region of Ethiopia, is that the animals are free to roam; they come into the homes on a voluntary basis, rather than being forced to do so. Why? Baynes-Rock suggests, and in this I am convinced by his arguments, that this is an alliance of humans and domesticates fostered by the presence of a threat of predators. .[42]

Close interspecies entanglements, illustrated by the bodily communication and exchange between humans and horses, defy any notion of a clear separation of humans and other creatures in building even the human moral landscape in specific settings. It is the *particular* interaction with *specific* others that is important in shaping our identity, including those others from other species groups. Specific virtues and vices are played out in our relationships with these others, and therefore, just as stuffed animals become the sounding board in the moral development of children, so real and specific animals of a specific kind do more than this—for they show us how to live well and with others unlike ourselves. In spite of differences between species, my own experiences with horses has, without question, helped to shape who I am today and the values I hold. Recognizing the source of that guidance is a theological task as much as an ethical task, because it acknowledges the other creatures as agents with whom we share our common home.

Finally, it is through learning specific tasks in relationship with (in this case) horses that we begin to learn the art of practical wisdom, or prudential wisdom in human affairs. In this sense our interaction with horses leads to an emergent wisdom, one that is grounded in our common experiences with these others. How might these relationships serve to guide our decisions? My proposal is that in the case of domesticated animals like horses, with whom we have shared a common life and whose intelligence and personality has become well known to us, these bonds will be analogous to our relationships with other family members. We learn to behave well in and through our relationships with members of our community. That human-horse relationships are remarkably resilient, common experiences among those of very different cultural contexts shows the depth and strength of the bonds, and the potential for abuse. Interspecies ethics is not so much a prescription for how to act as a reminder to take our relationships with these others into account in our ethical decision-making. Interspecies ethics encourages less an extension of our world into theirs (e.g., through rights discourse), and more an empathetic entry into the lives of these domesticated others who have, through their close contact with us, become attuned to the cues and signals arising in human histories. Domestication need not just be about control or abuse, for the early evolution of domestication was

42. Ibid.

just as likely to have arisen because of a mutual fear among people and the domesticates of common predators. The Oromo horses are both domesticated and wild—free, on the whole, to associate with their owners or not. They and their human associates have an important lesson to teach us about entangled human-horse histories.

Overall, by way of conclusion, such encounters with horses point to a different constructive theological way of conceiving the human that is less about our differences from animals and other living creatures, and more about our relationships with them and, further, navigating those boundaries in positive rather than negative ways. This liminal space between creatures also opens up a particular way of considering who we are as humans in relation to God, namely, shared creatures in a theodrama. This play works to a script that is more often than not disguised from our eyes, and may seem improvised, but could well be detectable by sentient creaturely others. Humans are uniquely endowed with powers of language, as well as powers of cruelty, but that does not mean that other forms of encounter, communication, and interrelationship are insignificant.

For the best part of our evolutionary history as *Homo* communication was through gestures and other than verbal language. Deep evolution reinforces a theological anthropology that is suitably chastised with respect to claims for human supremacy or human exceptionalism. In fact, as I have indicated in this chapter, our moral awareness and compass has expanded through our relationships with others, culminating in our sense, at least for religious believers, that our relationship with God has primacy over other relationships in setting forth the moral task of humanity. But the primacy of that particular relationship with the divine should not so much work against our responsibilities in relation to other kinds but rather foster our sensitivity and awareness of shared and entangled creaturehood.

If we look into the eyes of a horse with openness and spend time with him/her, then we will see displays of character that reminds us clearly of our own moral worlds. That, as much as philosophical arguments, can inspire our specific responsibility to care in accordance with what we understand are the particular needs from a horse's point of view. We might not be able to get into the mind of another animal, but we can, especially in the case of domesticated ones, connect with his or her soul.

Bibliography

Baynes-Rock, Marcus. "Interview with T. Teressa on Oromo Horses." Recorded on April 16, 2016.

———. "In the Minds of Others" In *The Evolution of Human Wisdom*, edited by Celia Deane-Drummond and Agustin Fuentes, 47–67. Lanham: Lexington, 2017.

Baynes-Rock, Marcus, and Tigist Teressa. "Shared Identity of Horses and Men in Oromo, Ethiopia." *Society and Animals* (forthcoming).

Brandt, Keri. "A Language of Their Own: An Interactionist Approach to Human-Horse Communication." *Society and Animals* 12.4 (2004) 299–316.

Briard, Léa, Camille Dorn, and Odile Petit. "Personality and Affinities Play a Key Role in the Organization of Collective Movements in a Group of Domesticated Horses." *Ethology* 121 (2015) 1–15.

Chamove, Arnold S., Ocean J.E. Crawley-Hartrick, and Kevin J. Stafford. "Horse Reactions to Human Attitudes and Behavior." *Anthrozoös* 15.4 (2002) 323–31.

Davis, Dona, Anita Maurstad, and Sarah Cowles. "'Riding Up Forested Mountain Sides, In Wide Open Spaces, and Without Walls': Developing An Ecology of Human-Horse Relationships." *Humanimalia: A Journal of Humananimal Interface Studies* 4.2 (2013) 54–83.

Deane-Drummond, Celia. "Deep History, Amnesia and Animal Ethics: A Case for Inter-Morality." *Perspectives on Science and Christian Faith* 67.4 (2015) 1–9

———. "Natural Law Revisited: Wild Justice and Human Obligations to Other Animals." *Journal for the Society for Christian Ethics* 35.2 (2015) 159–73.

———. *The Wisdom of the Liminal*. Grand Rapids: Eerdmans, 2014.

Deane-Drummond, Celia, Agustín Fuentes, and Neil Arner. "The Evolution of Morality: Three Perspectives." *Philosophy, Theology and the Sciences* (October 2016) in press.

Guo, Kun, Kerstin Meints, Charlotte Hall, Sophie Hall, and Daniel Mills. "Left Gaze Bias in Humans, Rhesus Monkeys and Domestic Dogs." *Animal Cognition* 12 (2009) 409–18.

Hearne, Vicki. *Adam's Task: Calling Animals by Name*. New York: Skyhorse, 2007.

Ingold, Tim. "Prospect." In *Biosocial Becomings. Integrating Social and Biological Anthropology*, edited by Tim Ingold and Gisli Palsson, 1–13. Cambridge: Cambridge University Press, 2013.

Lawrence, Elizabeth. *Rodeo: An Anthropologist Looks at the Wild and the Tame*. Knoxville, TN: University of Tennessee Press, 1982.

Ethiopian Television. "Ethiopians in traditional Oromo costume pay their respects to PM Meles Zenawi." Ethiopian Television on DireTube. Posted August 31, 2012. https://www.diretube.com/ethiopian-news-ethiopians-in-traditional-oromo-costume-pay-their-respects-to-pm-meles-zenawi-video_167eae118.html.

Racca, A., K. Guo, K. Meints, and D. S. Mills. "Reading Faces: Differential Lateral Gaze Bias in Processing Canine and Human Facial Expressions in Dogs and 4-year old Children." *PLoS ONE* 7 (2012). doi:10.1371/journal.pone.0036076.

Smith, Amy Victoria, Leanne Proops, Kate Grounds, Jennifer Wathan, and Karen McComb. "Functionally Relevant Responses to Human Facial Expressions of Emotions in the Domestic Horse (*Equus caballus*)." *Biology Letters* 12 (2015). http://dx.doi.org.10.1098/rsbl.2015.0907.

Willett, Cynthia. *Interspecies Ethics*. New York: Columbia University Press, 2014.

Wolfe, Cary. *Before the Law: Humans and Other Animals in Biopolitical Frame.* Chicago: University of Chicago Press, 2012.

Wynne, Clive D. *Animal Cognition: The Mental Lives of Animals.* Basingstoke: Palgrave McMillan, 2001.

Culture of Sacrifice

Another Meditation on Domestication in the 21st-Century—US Anthropocene

—LAURA HOBGOOD—

Jesus loves me, He who died
Heaven's doors to open wide,
And to wash away my sin
Let the little child come in.[1]

Singing "Jesus Loves Me" as a child is standard practice in churches throughout the United States. Its very mundane and non-threatening sweetness pronounces a religion that is benign, always loving, and, obviously, welcoming. But when one thinks deeply about its salvific claims, ones that are ingrained in many Christian children so early on in their lives, some important questions arise. And these questions, along with the answers drawn, impact not only humans, but all beings living on Earth. The questions delve into atonement, sacrifice, salvation, and hierarchies, just to name a few of the complicated relationships. What does it mean for God

1. This verse from the children's hymn "Jesus Loves Me," was originally part of a poem written by Anna B. Warner in 1860. There are various versions of the wording for the verse, but the one quoted here is relatively common and popular. For more information on the history of this poem/hymn see McElrath, "The Hymnbook as a Compendium of Theology," and Reynolds, "Women hymn writers and hymn tune composers in the Baptist Hymnal, 1991."

to die for "me" ("us"), for humans (alone)? What does this then say about all of the other beings on the planet? But most importantly, what does the sacrifice of God for humans say about who humans think we are?

There's a building on the campus where I teach that is always difficult for me to enter. This particular building houses most of the laboratories where research is conducted on animals, primarily on rats. Upon stepping over the threshold of that building a sense of suffering and sadness envelopes me. These amazing, intelligent subjects of their own lives (rats) are being used in a variety of experiments in order to potentially better or enhance or extend the lives of humans (more on that later as, arguably, only some privileged humans benefit). These experiments do not better the lives of rats—that is not the intended purpose.[2]

Maybe one of the reasons that this place of experiment and suffering strikes me as so tragic is because of a childhood memory. A family of rats came home with me from a biology classroom over spring break when I was in middle school. Someone had to take care of them after all, right? And that was the case even if that someone had failed to ask for her parents' approval. My compassionate mother, always accepting of animals in the house (and usually dealing with them more than whoever brought the new companion home) was even a bit taken back by the presence of the mom rat with her litter of babes. My mom kept looking at the rat mom's tail saying it was just "creepy." Of course, we have no idea what the rat mom thought of my mother, though I assume she found my mom appealing after she became their primary caretaker for this brief period of time. I do recall, clearly, watching the mom rat gently care for her young (and learning from my mother as she gently cared for the rats). Mom rat also asked the humans around her for attention and loved to be held and caressed; she also loved a belly rub and a tickle (rats do laugh).[3] The irony of using rats for so much behavioral research for pharmaceuticals and other purposes is that we know a lot about rats. We know they are complex animals who can even read the pain in the faces of other rats and display empathy.[4]

But, as "Jesus Loves Me" and as rats, arguably, remind us in real, frequently suffering, flesh, sacrificial atonement, and all of its ramifications for bodies who are sacrificed, abounds and usually wins the day. Just what is

2. I think it is important to note here that many scientists do the best they can to make sure that the animals in their laboratories, including rats, have the best possible lives while they are in their labs and they follow IACUC (Institutional Animal Care and Use Committee) standards to the letter.

3. See Underwood, "Watch these ticklish rats laugh and jump for joy."

4. There is extensive research on rats and their emotional lives. For a concise overview with references included see Bekoff, "The Emotional Lives of Rats."

this theological assertion anyway? Sacrificial atonement theology declares that Jesus (God) died (was sacrificed) to save humans. Humans did not have the moral or historical capacity to do so for themselves, so God decided to die for us. It is a stunning claim, one that carries incredible power on so many, and myriad, levels. But in some ways the most chilling claim, while also the most beautiful in some humans' estimation, is that humans, and humans alone, are worthy of even the death of God. Humans apparently require the death of God for our salvation and God willingly dies for us. The next step in this equation, then, is potentially disastrous, particularly for other animal species. If God willingly dies for us then what comes next? So, we ask, why should not all other beings also die for us? Are we not worthy of the sacrifice of everyone and everything? Of course we are: God died for us humans, at least according to traditional (orthodox) Christian atonement theology. These theological ideas, and the implications they have for other animals, are the topic with which this essay grapples. What does it mean to live in a culture of sacrifice that exists for the benefit of just one species?

Humans entered the twenty-first (Christian) century with a bang, both literal and symbolic. Not only did the human population quickly pass the seven billion mark with exponential population growth predicted and few immediate possibilities for slowing, but myriad other-than human animal populations were struggling for survival, with the exception of some domesticated animals. Fish, birds, mammals, amphibians, and insects (and of course many plants) are now all part of what many designate as the "sixth great extinction period."[5] And this period of extinction is, by all accounts, caused by humans and our impact on the planetary ecosystems, climate, and landscape. The vast majority of scientists suggest that no previous great extinction resulted from the direct impact of one other species on the entire global biosystem.[6] How did we achieve such a destructive status? Why are most humans thriving (at least the species undeniably is, certainly some individuals and communities are not), while most other species are losing the fight against extinction or against having any form of control over their own lives? What does this have to do with Christian theology? And, for the purposes of my thesis, one other question must be posed: Are humans willing to sacrifice so many other lives, eventually even our own if scientific predictions are true, in order to remain dominant for the near future?

5. Kolbert, *The Sixth Extinction*, is likely one of the best and most accessible studies of this period in Earth's history.

6. See, among others, Davies, *The Birth of the Anthropocene*, 35–38.

The Sacrifice Conundrum

Sacrifice is such a loaded and contextual word. At times, sacrifice means that which one denies oneself and can be viewed as a kind of virtue. When a soldier sacrifices her life in a battle, or a mother sacrifices her body in birth, or a monk sacrifices his caloric intake so others around him might eat, choices are made and life or limb or other losses are accepted, to a certain extent, willingly. These kinds of ascetic sacrificial practices are central to many religious systems. Indeed, it is difficult to imagine a religious system and historically improbable to locate one that does not have a sacrificial component in the ascetic, virtuous sense of the word. In Hinduism, wives sacrifice eating from moonrise to moonrise to sustain the strength of their husbands; in Christianity, female, male, and LGBTQ religious devotees (monks, nuns) sacrifice a primary human relationship to "marry Christ";[7] in Islam, during Ramadanall devout (and physically able) believers sacrifice food and beverage from sunrise to sunset.

In Christianity specifically there are ancient texts that provide the foundation for such sacrificial living. Athanasius's *Life of St. Anthony*, a text which became a model for later such hagiographical lives, tells of Anthony's transition into monasticism:

> Anthony went out immediately from the church and gave the possessions of his forefathers to the villagers—they were three hundred acres, productive and very fair—that they should no longer be a burden upon himself and his sister. And all the rest that was moveable he sold, and having got together much money he gave it to the poor.[8]

In many ways ascetics practiced a way to counterbalance the increasingly powerful and wealthy institution of the church. They were frequently referred to as living martyrs, taking on the witness of early Christians in the Roman Empire. Sacrificial ways of living such as these continue throughout the history of Christianity. A brief list of some of the more prominent figures includes (but is obviously not limited to): Benedict, Hildegard, Francis, Clare, Mother Teresa, and Martin Luther King Jr. Tens of thousands of Christians chose (and still choose) to live within such a sacrificial mind-set and practice. Many were even directly involved with living in a way that

7. Of course there is research that suggests not all of those who take vows of celibacy maintain them; there is also fascinating research that suggests some female religious communities were also safe spaces for females who were not heterosexual. There are a number of sources for this, including Gramick, *Homosexuality in the Priesthood and the Religious Life*, and Brown, *Immodest Acts*.

8. Athanasius, *Life of Saint Anthony*, 2.

recognized and took seriously the lives of other animals. Abstaining from the consumption of flesh was a common practice for both lay and monastic Christians through at least the Middle Ages. Women, in particular, focused on food consumption and abstaining from eating flesh.[9] In their fascinating study of Christian diets, Grummett and Muers point out that "[r]igorous and clearly defined dietary abstinence, especially from red meat, has been a key Christian discipline from a very early stage in the forging of Christian identity."[10] In the twenty-first century official groups of Christians who live a type of sacrificial life in relationship to animals, such as the Christian Vegetarian Association, continue in these ancient traditions.[11]

But sacrifice also means offering up someone, most often at least in its classic construction someone other than oneself, to the divine in order to gain a favor from the divine. The roots of the understandings of sacrifice in Christianity are found primarily in the Hebrew Bible. It seems important to place the idea of animal sacrifice in the context of overall diet as referenced in the earliest sections of the Bible. The first nine chapters of Genesis describe a life that is, primarily, vegetarian. Only the green plants are given to humans for food. After the Noah flood stories humans are essentially cursed: "the fear and the dread of you shall rest on every animal of the earth, and on every bird of the air, on everything that creeps on the ground, and on all the fish of the sea; into your hand they are delivered" (Gen 9:2). God gives humans permission to eat animals, with certain restrictions for sure. But the relationship changed.

In terms of animal (and sometimes human) sacrifice specifically, it enters stories early in the Hebrew Bible. In Genesis 4 Abel sacrifices a sheep to God while Cain offers fruits. Abel's sacrifice of an animal was regarded well, while Cain's offering of vegetation was not. This sacrifice of a sheep leads to the first human murder in the Hebrew Bible (when Cain kills Abel). The other foundational story of sacrifice is God's call to Abraham to sacrifice his son, Isaac (Gen 22). Abraham is prepared to do so, literally on the verge of doing so, when God stops him and provides a ram as a burnt offering in place of Isaac.[12] Many theologians argue that this story, along with texts

9. See Bynum, *Holy Feast and Holy Fast.*

10. Grummet and Muers, *Theology on the Menu*, 35.

11. See http://christianveg.org/.

12. In the book of Judges, chapter 11, Jephthah is required by God to sacrifice his (unnamed) daughter; no other animal is given by God as a substitute and the daughter is, indeed, sacrificed. There is definitely a gendered connection to sacrifice, but that is another paper. For an interesting interpretation of this see Trible, *Texts of Terror*.

from the prophet Isaiah, is a model upon which the Christian ideas of the sacrifice of Jesus, God's son, is based.[13]

After the first Temple was built in Jerusalem, sacrifices of animals became a central aspect of worship and the Temple became the only location where sacrificial rituals could be performed.[14] Most sacrifices described in the Hebrew Bible, as well as those performed at the Temple in Jerusalem, involved domesticated animals, specifically sheep, goats, and cattle (Lev 1:2). A combination of factors contributed to this idea in ancient Israel that the perfect sacrifice is one that involves domesticated animals. In his fascinating study, Jonathan Klawans points to several of these influences including ritual purity and the imitation of God as shepherd in the sanctuary. He suggests, "in ancient Israel, sacrifice involves—in part—the controlled exercise of complete power over an animal's life and death. This is precisely one of the powers that Israel's God exercises over human beings: 'The Lord kills and brings to life' (1 Samuel 2:6, cf. Deuteronomy 32:39)."[15] So, as Robert Daly points out, the "sacrificial ritual was the center of the religious life of ancient Israel."[16]

This particular religious context, along with the broader context of animal sacrifice in the Mediterranean world, specifically in the religious sensibility of the Roman Empire as a whole, set the stage for early Christianity. Indeed, as stated by Gilhus, "sacrifice was the universal language of worship in the ancient world."[17] Sacrifice had powerful positive connotations associated with offering something pleasing to a deity:

> In the ancient world sacrifice . . . was viewed as a cultic act, a religious ritual. . . . [S]acrifices were preferably as large as possible and were always offered to a god, thus indicating a recognition of the deity's superior status . . . . [S]acrifices were generally performed joyously as part of a public holiday or celebration and were often associated with sentiments of petition and thanksgiving.[18]

Not only in ancient Israelite Temple religion, but also in the Roman Empire as a whole the idea of sacrificing domesticated animals took hold as a central premise. At major religious festivals throughout the Empire a "proper sacrifice was one in which an unblemished animal was induced to suggest

13. For an outline of the images of Jesus as a sacrificial lamb, see Perlo, *Kinship and Killing*, 83–85.

14. Daly, "The Power of Sacrifice in Ancient Judaism and Christianity," 187.

15. Klawans, "Sacrifice in Ancient Israel," 68–69.

16. Daly, "The Power of Sacrifice in Ancient Judaism and Christianity," 181.

17. Gilhus, *Animals, Gods and Humans*, 157.

18. Daly, "The Power of Sacrifice in Ancient Judaism and Christianity," 183.

its willingness by stretching forward its neck."[19] The sacrificial lamb must cooperate with her demise.

When Christianity became a legal religion, under Emperor Constantine in 313 CE, then the only official and legal religion, under Emperor Theodosius, it became increasingly urgent to distinguish Christians from other religious practitioners, specifically Jews and "pagans" (those who practiced the old Roman religions). One of the ways early Christians sought to differentiate themselves from both Roman and Jewish practices was to eliminate animal sacrifice from their rituals. This might have been a forced distinction as it is apparent that some Christians were still ritually sacrificing animals. Indeed, some Christian communities (mostly in the Mediterranean world) still practice animal sacrifice or symbolic versions of it into the present day.[20] But to serve the political needs of the empire, Theodosius, whose code issued in 381 privileged Christianity over all other religions in the empire, banned animal sacrifice.

From that point forward "Christians worshipped a god who was himself the sacrifice."[21] The tide turned and Christianity accepted a theological position focused on God (in the form of the Son, Jesus) as the ultimate sacrifice. He became the Lamb of God who takes away the sins of the world. God died for humans.

Humans Worthy of the Sacrifice of God?

So what does it mean when a species, homo sapiens, decides that the divine, the one God, determined to offer up itself to itself for our salvation? Might it tell humans that if even God is willing to sacrifice life for us, should not all beings be on the altar for human salvation? In this cosmic worldview all beings, even the planet Earth itself, are willing and even yearning to sacrifice for humanity. Remember, from the models of the ancient Mediterranean, the unblemished sacrifice offered his/her own neck up to the priest.

Millions of domesticated animals in contemporary US culture die each year for various forms of human consumption. Does the acceptance of the sacrifice of millions of animals fit into the framework of sacrificial atonement theory that is central to various forms of Christian theology? Or, more

19. Kyle, *Spectacles of Death in Ancient Rome*, 35.

20. See Grummett and Muers, *Theology on the Menu*, 109–15 for examples of continuing animal sacrifice in Christianity. See Hobgood, *Holy Dogs and Asses*, introduction, for an example of a symbolic ritual sacrifice in contemporary Christianity.

21. Gilhus, *Animals, Gods and Humans*, 157.

poignantly, does Christian sacrificial atonement theology create an ethos wherein such a massive level of death of domesticated animals is acceptable? Indeed, there are also millions of non-domesticated animals sacrificed to make way for and benefit the (generally luxurious) desires of humans. One need only think of those animals whose habitats are destroyed because of the rapid growth of the human population and the use of natural resources at unprecedented rates or of those animals who are killed by big game trophy hunters to see that the sacrifice of other animals to humans is a global phenomenon that does not only involve domesticated animals. But, following in the steps of Jonathan Z. Smith, traditional Mediterranean religious-based sacrifice is a "meditation on domestication." While Smith was referring primarily to agriculturalist and pastoralist societies in his seminal essay "The Domestication of Sacrifice," there seems to be a significant link to the ways in which massive numbers of animals in the agricultural and scientific systems are sacrificed in the US today. Does this connect to Christian theology which dominated and dominates the religious sensibility of the US?

As explained by Karl Barth, one of the most influential Christian theologians of the twentieth century, sacrificial atonement is:

> What took place is that the Son of God fulfilled the righteous judgement on us human beings by himself taking our place as a human being, and in our place undergoing the judgement under which we had passed. . . . Why did God become a human being? So that God as a human being might do and accomplish and achieve and complete all this for us wrongdoers, in order that in this way there might be brought about by him our reconciliation with him, and our conversion to him.[22]

Some scholars argue that when humans actively engaged in animal sacrifice as part of acknowledged religious practices, that animals were, in some ways, much more elevated and, at the very least, recognized as actors to a certain extent. For example, Klawans, in the study of animal sacrifice in ancient Israel mentioned above, states that:

> If sacrifice poses an ethical problem, it is because animals are killed in the process. Yet the (incomplete) elimination of ritual sacrifice from (most) contemporary religious practice has done no good for the animals. Whoever feels smug about the elimination of sacrificial altars can just visit a slaughterhouse or a laboratory: neither is a more welcome place for an animal than an ancient temple.[23]

22. Barth, quoted in McGrath, *Christian Theology*, 354

23. Klawans, "Sacrifice in Ancient Israel," 65.

So, in the examples given in the historical outline above, animals mattered! Do they matter less now that they are sacrificed in large, unseen hoards?

Many of the contemporary sacrifices are also based on the current human interest in, dependence on, and hope for our technological prowess. That prowess continues to reinforce the idea that we are not only worthy of God's sacrifice for our salvation, but that we are almost capable of bringing it about ourselves through our supposed brilliance, through our exceptionalism. In her most recent book, Donna Haraway describes the horrors of the Anthropocene in relationship to technology:

> The first is easy to describe and, I think, dismiss, namely a comic faith in technofixes, whether secular or religious: technology will somehow come to the rescue of its naughty but very clever children, or what amounts to the same thing, God will come to the rescue of his disobedient but ever hopeful children.[24]

So what to do in the twenty-first century ethos of animal sacrifice when God decides to die for humans? Do other animals matter at all? What is their situation?

Animal Sacrifices in the Twenty-First Century

At this point in human history, cultural practices reflect a decision on the part of humans to act as the dominant, most worthy, and most important species that has ever lived on planet Earth. With that in mind, all other animals are subject to our needs and wants. As mentioned early in this essay, rats can live a short and frequently painful life. That reality is acceptable in the current cultural world as long as some possible medical or behavioral "progress," a word I use problematically, related to the lives of humans is gained (or maybe not even gained; experiments often have no positive outcomes). And scientific laboratories are not the only locations of sacrifice. Factory farms, destroyed habitats, and animal control facilities all participate in the sacrifice of other animals to fulfill human desires. Just how many animals lives end in a sacrifice to human consumption (a word I use deliberately here since even medical procedures are a form of human consumption)? The numbers are staggering and are shared below are just for those used in research (medical, scientific in general, product/industry) and as food.

The US Animal Welfare Act regulates the reporting of figures for animals used in research, but it does not cover rats, mice, birds, or fish. Rats

24. Haraway, *Staying with the Trouble*, 3.

and mice are used in numbers that far exceed those of all other animals in research combined, so it seems important to at least provide estimates of those numbers, which I do below. According to the US Department of Agriculture, the governing body to whom these statistics are reported, the following numbers were recorded for fiscal year 2015:

Animal	Number Used in Research
Cats	19,932
Dogs	61,101
Guinea Pigs	172,864
Hamsters	98,420
Pigs	46,477
Primates (nonhuman)	61,950
Rabbits	138,348
Sheep	10,678
Other Farm Animals	27,786
Other Animals	130,066[A]

A. See "Research Facility Annual Summary Report," https://www.aphis.usda.gov/aphis/ourfocus/animalwelfare/SA_Obtain_Research_Facility_Annual_Report, last accessed January 30, 2017.

That amounts to over 750,000 animals used in experimental laboratories in the year 2015 in the United States. But, as mentioned above, that is not the full story. Since rats and mice are not reported in these official statistics, it is difficult to know how many are in use. However, the figures for rats and mice are reported in the European Union and in the EU those two categories comprise over 93 percent of animals used in research.[25] If the same is true in the United States, and it is most likely at least close in terms of comparative numbers, then one can assume that over eleven million rats and mice are used for research annually in the US.

In a complex, but very important, chapter regarding animal use in medical experimentation in her book *When Species Meet*, Donna Haraway offers an important analysis of animals as "sacrifices" in medical laboratories. At one point she responds to Jacques Derrida's philosophical struggle with the human-animal divide, what he refers to as an "unbridgeable gap" in which "lies

25. See "Seventh Report on the Statistics on the Number of Animals used for Experimental and other Scientific Purposes in the Member States of the European Union," http://eur-lex.europa.eu/resource.html?uri=cellar:e99d2a56-32fc-4f60-ad69-61ead7e377e8.0001.03/DOC_1&format=PDF, last accessed January 30, 2017.

the logic of sacrifice" and "within which there is no responsibility toward the living world other than the human."[26] In this religious and secular worldview only the human can be murdered, but it is allowable to kill animals. Thus, "animals are sacrificed precisely because they can be killed. . . . [Sacrifice] works; there is a whole world of those who can be killed, because finally they are only something, not somebody, close enough to 'being' in order to be a model, substitute, sufficiently self-similar . . . but not close enough to compel response."[27] Haraway replies to these ideas of Derrida's with some admitted hesitation. She contends that it is a "misstep to separate the world's beings into those who may be killed and those who may not and a misstep to pretend to live outside killing."[28] She goes on to suggest that '[p]erhaps the commandment should read, 'Thou shalt not make killable.'"[29]

Animals who are sacrificed in laboratories are "killable" by all accounts in dominant US culture. But they are not the only ones who are killed in vast numbers. Even more pervasive is the sacrifice of millions of domesticated animals for food. According to the US Department of Agriculture in January 2017, so these are one-month totals, these animals were killed for food in the US:

Animal	Number Slaughtered[A]
Cows	2,580,000
Calves (veal)	46,600
Pigs	10,100,000
Sheep	177,000
Chickens	760,363,000
Turkeys	19,541,000
Ducks	2,492,000

A. See "Poultry Slaughter," http://usda.mannlib.cornell.edu/MannUsda/viewDocumentInfo.do?documentID=1131 for chickens, turkeys, ducks, report filed February 24, 2017; see "Livestock Slaughter," http://usda.mannlib.cornell.edu/MannUsda/viewDocumentInfo.do?documentID=1096 for cattle, calves, pigs and sheep, report filed February 23, 2017—both last accessed March 5, 2017.

26. Haraway includes an extensive footnote with references to both Derrida's work (with Jean-Luc Nancy) "'Eating Well: or the Calculation of the Subject': An Interview with Jacques Derrida," and with the myriad responses to it. Haraway, *When Species Meet*, 334.

27. Ibid., 78–79.

28. Ibid., 79.

29. Ibid., 80.

The Humane Society of the United States estimates that 9.2 billion animals were killed for food in the United States in 2015.[30] While that number might seem staggering, it should be noted that this does not include all animals slaughtered, only those for whom reporting is required. Missing animals are all fish, crustaceans, rabbits, buffalo, and more. If one simply counts that number of fish estimated for the US alone (approximately 500 million annually), the total number of animals killed for food tops 10 billion.[31]

Of course, there are other ways animals are sacrificed in addition to scientific experimentation and for food consumption. Some are killed in animal shelters because they have no homes (dogs and cats primarily), others are killed for sports such as hunting or dog fighting, others are hit by cars, still others lose their habitat. How does all of this connect to the theology of sacrifice that is foundational to most forms of Christianity?

While the US is not officially a Christian nation, certainly Christianity has been the dominant cultural religious force since Europeans invaded North and South America. Early in this process the concept of manifest destiny, derived from Christian and Eurocentric ideas, became a reality. European settlers expanded rapidly across the continent, carrying Christianity (primarily, certainly other religious traditions as well) and its ideas and theology with them. Indigenous people were sacrificed, wild places were sacrificed, buffalo were almost driven to extinction (in part to destroy indigenous people). And, today, billions of animals are sacrificed annually.

The theological suggestion I am making here is that Christianity, with its idea that humans are important enough for even God to sacrifice Godself to us for our salvation, lays the foundation for all other beings to sacrifice themselves to humans as well. We are, indeed, that worthy. Rarely do we even consider the suffering, it is rendered invisible. Rarely do we consider the ethical implications of the vast numbers of animals who are dying. As a matter of fact, we tend to deny that there are ethical implications at all. Christianity laid the foundation for a culture of sacrifice and perpetuated that idea, all animal offerings placed on our altar.

Is this theological-cultural suggestion a stretch? One could of course argue that animals are killed globally in cultures where Christianity is not the dominant religious tradition. Certainly that is the case. But the trajectory of contemporary scientific experimentation on animals, which was at times quite gratuitous (one need only read some of Descartes's accounts of

30. See the Humane Society of the United States, "Farm Animal Statistics: Slaughter Totals" at http://www.humanesociety.org/news/resources/research/stats_slaughter_totals.html??referrer=https://www.google.com/. Last accessed March 4, 2017.

31. For fish numbers see "Fish Count Estimates," http://fishcount.org.uk/fish-count-estimates. Last accessed March 4, 2017.

experimentation to determine this), began in the world of European Christianity where animals were classified as machines. And a diet that includes animal protein at the levels of US consumption is an anomaly in most of human history. Plant-based diets were the norm for almost all cultures (with the exception of some cultures in climates that prohibited plant-based diets, such as people living in the Arctic Circle or for some cultures who survived on water-based animals). And, still today, many cultures focus on plant-based diets. But the US is rapidly exporting an animal-protein–based diet. For example, in 1982 the average person in China consumed thirteen kilograms of meat per year. But 2016 that amount increased to sixty-three kilograms per year. Interestingly, China is now in the midst of a public health and environmental (production of meat is one of the major contributors to climate change) campaign to reduce meat consumption by 50 percent.[32]

So the question remains, does the idea of sacrificial atonement, a central component of Christianity, allow for a relatively easy justification for the death of other-than-human animals for the benefit of humans? And, in some cases, not even a benefit but a luxury. Ingesting an overabundance of animal protein is, arguably, detrimental for human health. One only needs to ask if veal is a necessary sacrifice or simply a cruel luxury. Yet, if even God will die for us, certainly the calf who has never been in the grass or seen the sun, but has lived life in a tiny crate should die for humans as well.

In a time when other-than-human animals' lives are under threat from so many directions and when the human population is exploding, it is incumbent upon humans to rethink theological positions and religious convictions that put other animals at even more risk. Christianity, a tradition that has historically offered such a powerful voice on social justice issues for humans, has the resources at hand to become a religion that is compassionate toward other animals as well. While there are few references to connections with other animals in the canonical Christian texts, there are hints about the possibilities. Luke 14:5 tells the reader that the Sabbath laws can be broken to save an ox, Mark 1:13 tells the reader that Jesus was "with the wild animals" while he was resisting temptation. And in extracanonical texts, stories of Jesus and animals are prevalent. He saves a lioness and her cubs, he resurrects a salted fish, he admonishes a man who is beating his mule.[33] The current state of human relationships with and power over other-than-human animals calls us to rethink our anthropocentric traditions.

32. See Milman and Leavenworth, "China's plan to cut meat consumption by 50% cheered by climate campaigners."

33. See Hobgood, *Holy Dogs and Asses*, chapter 3.

Reconsidering the Christian theology of sacrifice in a way that decenters humans might be a step in this direction.

Bibliography

Athanasius. *Life of Saint Anthony*. In *A Select Library of Nicene and Post-Nicene Fathers of the Christian Church: Second Series Vol. 4*, edited by Philip Schaff and Henry Wace, 188–221. New York: Scribner, 1900.

Bekoff, Marc. "The Emotional Lives of Rats: Rats Read Pain in Others' Faces." *Psychology Today*, April 1, 2015. Accessed April 13, 2018. https://www.psychologytoday.com/us/blog/animal-emotions/201504/the-emotional-lives-rats-rats-read-pain-in-others-faces.

Brown, Judith C. *Immodest Acts: The Life of a Lesbian Nun in Renaissance Italy*. Oxford: Oxford University Press, 1986.

Bynum, Caroline Walker. *Holy Feast and Holy Fast: The Religious Significance of Food to Medieval Women*. Berkeley, CA: University of California Press, 1989.

Daly, Robert. "The Power of Sacrifice in Ancient Judaism and Christianity." *Journal of Ritual Studies* 4.2 (1990) 181–98.

Davies, Jeremy. *The Birth of the Anthropocene*. Oakland, CA: University of California Press, 2016.

Derrida, Jacques. "'Eating Well,' or the Calculation of the Subject: An interview with Jacques Derrida." In *Who Comes After the Subject?*, edited by Eduardo Cadava, Peter Connor, and Jean-Luc Nancy, 96–119. New York: Routledge, 1991.

Gilhus, Ingvild Saelid. *Animals, Gods and Humans: Changing Attitudes to Animals in Greek, Roman and Early Christian Ideas*. New York: Routledge, 2006.

Gramick, Jeannine, ed. *Homosexuality in the Priesthood and the Religious Life*. New York: Crossroad, 1989.

Grummett, David, and Rachel Muers. *Theology on the Menu: Asceticism, Meat and Christian Diet*. New York: Routledge, 2010.

Haraway, Donna. *Staying with the Trouble: Making Kin in the Chthulucene*. Durham, NC: Duke University Press, 2016.

———. *When Species Meet*. Minneapolis: University of Minnesota Press, 2008.

Hobgood, Laura. *Holy Dogs and Asses: Animals in the Christian Tradition*. Urbana, IL: University of Illinois Press, 2008.

Klawans, Jonathan. "Sacrifice in Ancient Israel: Pure Bodies, Domesticated Animals and the Divine Shepherd." In *A Communion of Subjects*, edited by Paul Waldau and Kimberly C. Patton, 65–80. New York: Columbia University Press, 2006.

Kolbert, Elizabeth. *The Sixth Extinction: An Unnatural History*. New York: Henry Holt, 2015.

Kyle, Donald. *Spectacles of Death in Ancient Rome*. New York: Routledge, 2012.

McElrath, Hugh T. "The Hymnbook as a Compendium of Theology." *Review and Expositor* 87.1 (February 1990) 11–31.

McGrath, Alister E. *Christian Theology: An Introduction*. Oxford: Blakewell, 1994.

Milman, Oliver, and Stuart Leavenworth. "China's plan to cut meat consumption by 50% cheered by climate campaigners." *The Guardian*, June 20, 2016. Accessed

April 17, 2017. https://www.theguardian.com/world/2016/jun/20/chinas-meat-consumption-climate-change.

Perlo, Katherine Wills. *Kinship and Killing: The Animal in World Religions.* New York: Columbia University Press, 2009.

Reynolds, William J. "Women Hymn Writers and Hymn Tune Composers in the Baptist Hymnal, 1991." *Baptist History & Heritage* 41.1 (Winter 2006) 114–118.

Smith, Jonathan Z. "The Domestication of Sacrifice." In *Relating Religion: Essays in the Study of Religion*, 145–59. Chicago: University of Chicago Press, 2004.

Trible, Phyllis. *Texts of Terror: Literary-Feminist Readings of Biblical Narratives.* Philadelphia: Fortress, 1984.

Underwood, Emily. "Watch these ticklish rats laugh and jump for joy." *Science* November 10, 2016. Accessed April 13, 2018. http://www.sciencemag.org/news/2016/11/watch-these-ticklish-rats-laugh-and-jump-joy.

Becoming Human within the Ecological Community

Encountering Life on the Farm with Thomas Berry and Edward Schillebeeckx

—ABIGAIL LOFTE—

Reinventing the Human

My awareness of participating in the Earth community begins as I was growing up on the Lofte family farm. Wildness exists alongside domestication in this space and connects me beyond my human family. Here arises a deep understanding of being part of a wider, planetary family—a communion of differentiated subjects that includes the human as one creature among others, all sharing the planet's past and future.[1] As one of many living here, I experience the effects of previous generations' decision-making about this land. I realize that my life impacts future generations, including generations of plants and animals in our more-than-human family, as we navigate that supposed fine line between wildness and tameness. We put up fences to mark borders between these worlds, though there really is no barrier; the wild and domestic run right into the other on the farm and form one community. Fences do little for keeping out animals like raccoon and geese, though they are effective for keeping in cattle and hogs. It is a complicated dance, letting the wild things be wild while still controlling what is necessary for sustaining our lives. A farm is a place of

1. Jenkins, *Ecologies of Grace*, 96–97.

controlled wildness; the wild grows where it will under the influence of the human. One of the land's own offspring controls the fate of it all.

The complexity of farm relationships, and my own place in this community, comprises a "fifth Gospel"—a personal story that I contribute to the narrative of wider Christian tradition. Considering our context, both throughout recorded history and evolutionary development, we cannot ignore that our emergence is interlaced with that of all other creatures. This living tradition secures a place for Earth's belonging, and for mine, as I reimagine a new path from destructive relationships toward one of celebration for creaturely particularity and community. With the addition of concerns about the relationship between humans and Earth, the fifth Gospel becomes the place where I retrieve history and create the possibility of a future world. Understanding existence in a larger context and sharing relationships with creatures around the globe, I remember our emergent history while recognizing our place in this collective context, and work toward securing a life-giving place in the story while making commitments to the wider Earth community for a hopeful future together.

The creative tension implicit in life on a farm is indicative of what any family must endure and navigate for survival. Essential to farm life is negotiating the priorities of providing for our human family while acknowledging the needs of our more-than-human family, the context in which we exist. Where wildness and domestication live side by side, decision-making affects creatures at every level as we strive to make a profit but also protect the Earth community that allows and enhances a profitable production. In order to achieve this balance, some destruction and violence is inevitable and necessary to the integrative wellness of the entire place—sometimes trees must be felled and animals killed for the safety and flourishing of other members. In this more-than-human family, I am united to this place at a haematological level as the nuances of the land's chemical composition literally course through my veins. This land is in my blood and we are entangled in a mesmerizing but complicated relationship. Coexisting with the rest of Earth in this piece of land, I am one of many creatures who live here navigating the complexities of relationships while trying to respect the needs of other members.

In this essay, I discuss how my relationships with domesticated land and life on the farm has influenced my ecotheology, directing me toward a reinvention of the human as a species in relationship with Earth. In order to make sense of such relationships, I draw upon the theology of Edward Schillebeeckx, who develops the notion of community with the wider world and a lifestyle commensurate with Gospel values, and the thought of Thomas Berry, who insists that we must reinvent the human at the species level if we are to live peacefully with Earth. Thus, I develop an ecotheological anthropology rooted in Earth's daily presence in our lives and that seeks to understand what it means to be in relationship with Earth. Such a reinvention of the human expands our understanding of community and moves communal boundaries beyond human-human and even human-animal relationships, embracing a broader human-planet relationship as an irreducibly plural, more-than-human family.

Schillebeeckx looks to the Christian community as the supreme example of what supportive relationships might look like. Beginning with the disciples who followed Jesus with a mission that endures today, the Christian community orients itself around Jesus' resurrection, which becomes the fulcrum to its development. Schillebeeckx's focus on the resurrection appeals to the centrality of experience and encounter for forming the disciples into a community of faith and enlivening the Christian movement after Jesus' life, death, and resurrection. When Schillebeeckx's insights about the resurrection's significance are applied to Earth issues, a "fifth Gospel" can be developed.[2] Schillebeeckx describes the "fifth Gospel" as our own personal stories that contribute to the larger narrative of Christian history. This develops a deeper meaning of the Christian community's tradition because by intertwining our stories with those of the gospel we become part of the living tradition that is active and developing throughout a collective history. Within our collective history all find a place of individual belonging in the present and the anticipation of a shared future concerned with the needs of others.

While Schillebeeckx's insight pertains to dialoguing the story of human relationships with tradition, Berry widens the focus to include the whole community of Earth relationships. Berry underscores the importance of expanding the community to include the other-than-human, emphasizing the importance of creatures existing as a communion of subjects rather than as a collection of objects.[3] The creative tension woven into the very fiber of the cosmos ensures that each creature cooperates with every other creature so that, working together in a symbiotic process, each becomes and is known through their relationships with others.[4] Responding to the needs of our time—or any time—must include concern for Earth because the bond shared by all things is essential to sustaining the creativity of the universe and keeps all creatures moving and developing together toward the far future.

2. Schillebeeckx, *God Among Us*, 127.

3. See, e.g., Berry, *Dream of the Earth* and *The Great Work*.

4. Swimme and Berry, *The Universe Story*, 219.

Community and Lifestyle

Navigating the tension of these relationships requires a shift in focus from human priorities to the needs of all planetary life. Such can only happen if the scope of concern widens from a narrow purview to a much broader vision and takes into account Earth as a dynamic community of life. Understanding that humans exist in the context of the wider world, the move away from a human-centered approach requires recognition of the continuity we share with Earth. Made from the same stuff and processes as other creatures, we necessarily emerge from a shared history, live in a shared present, and will share the same future. My history with our domesticated land—the history of a more-than-human family—is a complex story of emergence and becoming. We live in a unique network of relationships in which we are all intertwined for the support and sustainability of one another. As such, this community of life must be considered from more than just the human perspective or from a utilitarian framework.

The resurrection, for Schillebeeckx, provides the ideal lens by which to view a community. He asserts that, more than the raising of Jesus from the dead by God through the Spirit, the resurrection is *the* event responsible for the historical dimension of assembling a community of disciples and forming them in relationship. In doing so, they are prepared for the spiritual dimension of seeking a better future through a commitment to renewing the world.[5] To do this requires a life consonant with that of Jesus of Nazareth, who lived his life in cooperation with the kingdom of God. What unites Jesus' life to that of the disciples' is the Holy Spirit enlivening and continuing the work of believers while also acting on behalf of those in need. In solidarity with them, the early community of disciples becomes the church of today, where we continue to keep this story and tradition alive, remembering the Paschal Mystery of Christ through the ongoing process of becoming disciples ourselves through our participation in that same mission.

Paramount to this lifestyle is a reinterpretation of what it means to follow Christ so that the ministry of Jesus remains relevant for today. Schillebeeckx asserts that "responding to one's own new situations from out of an intense experience of God" instills in believers the impetus to do more than simply imitate Christ's life. Instead he suggests a move toward participation with his life, doing *as* he did, responding to the needs of our own time and to the needs of people arising out of present-day issues.[6] As the

5. Schillebeeckx, *Christ*, 642.

6. Ibid., 641.

church, we are impelled to pursue "practical-critical priorities," with a keen awareness of the sociohistorical and cultural challenges that cause distress for our communities.[7] Through a living witness and active participation in remembrance of the Paschal Mystery, our attention is focused upon seeking a future that enhances life and permits us, the community, to become the living, dynamic presence of Christ's activity still present in the world.

By participating with God's ongoing work in the present, the future remains open to us and to all who struggle and, despite challenges, seeking justice and reconciliation are valuable ways to keep open this possibility.[8] Through reflecting on God's work and then acting in cooperation, justice and love can be realized in ways that make healing and liberation possible, regardless of the reality of suffering and death. This commitment to responding to the needs of our time in an ecological context requires that integrity in relationships prevail and communication is authentic for seeking healing and relief for those who suffer. God's salvific work and the life of freedom offered through the ministry of Christ are the models for the community's efforts today for continuing this way of life, committed to the gospel.

The church becomes necessary for making this eschatological vision of a renewed future present in the world already, though incompletely, through collaboration between individuals and communities. As the place that regards the human experience as sacred and reflects on the intricacies and diversity of it, the church offers a way to understand human experience and the circumstances that shape it. Schillebeeckx calls experiences of suffering and injustice "negative contrast experiences" that occur throughout history and have deleterious effects.[9] The human experience of suffering assumes an initial desire for happiness, whereas the experience of injustice assumes the human capacity for integrity in relationships. For the sake of ameliorating the effects of suffering, humans look forward to a transformative, final future where suffering takes on ultimate meaning. This future is only achieved through the suffering we wish to avoid, which simultaneously opens forward the way to this future of transformation and, if it is to be believed, directs humans to live with more care and compassion for each other and also to work for the elimination of the causes of suffering and oppression. This way of life requires an intentional integration, beginning with the human, of disparate creatures for the planet's continued emergence. Eliminating suffering for creatures who struggle, the material world

7. Simon, "Salvation and Liberation in the Practical-Critical Soteriology of Schillebeeckx," 495.

8. Ibid., 496.

9. Schillebeeckx, "Naar een 'definitieve toekomst,'" 45–47, quoted in Schreiter, "Contrast Experiences," 54–56.

must be valued beyond its instrumentality. A collaboration of humans to ease planetary suffering, however, first requires a reframing of the human context into an Earth context and a commitment to seeing the Earth as self-referent and the human as derivative—a reimagined anthropology Thomas Berry calls "the reinvention of the human."[10]

The power of suffering for Earth and ourselves mobilizes us to act to seek a better future, to create a more humane humanity, while being critical of sociohistorical constructs that allow suffering.[11] It is our job to offer resistance to such suffering and exploitation so that evil is refused the right to exist. Yet, what is perhaps more necessary is for this resistance to be united with action and involvement for opposing the causes of suffering. While we cannot make sense of suffering, we can remember our own suffering and the power of memory compels us to work and act on behalf of others, and ourselves, for liberation that pursues our original desire for happiness and relationships of integrity.

We demonstrate the salvific power of God's love through orthopraxis where, in our daily living, we commit to loving others as God loves us. God's love does not discriminate, so in our loving there is also no room for distinction or division. Through his ministry, Jesus exemplifies this love to others and invites them to participate with him through engagement with the world, not withdraw from it.[12] Through conversion to this way of life that meets resistance with action, we arrive at renewed belief and involvement that are cultivated through prayerful discernment and based upon a faith in God's coming kingdom. This conversion is also required of humans so that, at a species level, our relations with Earth will enhance life, promoting creativity instead of interrupting it.[13] Our assumption of more and more power over Earth has come at the price of the intimate relationship we might otherwise have with Earth. We are no longer a communion of subjects, but a transcendent species controlling a collection of objects.[14] The

10. Berry, *The Great Work*, 18.

11. Schillebeeckx, *Christ*, 724–28.

12. Phillips, "Seeing with the Eyes of Faith," 248–49.

13. Note that this does not mean an absence of violence. Creative tension necessitates creative violence, but not violence that allows one species to dominate the others. Swimme and Berry assert it was the inception of the Cartesian mechanism that divorced the spiritual from the physical, reducing individuals to their parts rather than their integral whole. Through the promotion of this scientific method of inquiry, our skills for relating well to each other as a communion of subjects and our capacity for recognizing the universe's numinous qualities were stunted, leading to our inability to easily acknowledge the subjective identity of the created world and, therefore, ourselves within it. Swimme and Berry, *The Universe Story*, 227–31.

14. Ibid., 249.

lifestyle of the kingdom that Schillebeeckx promotes requires us to seek a better future for the whole Earth community through our interactions with other members as we commit to living like Jesus, who showed readiness to suffer with and for others in solidarity. As we join in this call to solidarity with our Earth-other-neighbors, the church becomes the witness to the transformative power of the gospel that seeks salvation and transformation for all who are oppressed and struggling.[15] Uniting ourselves to Christ's mission, we discover lives full of meaning and happiness because we have taken as our personal responsibility the well-being and safety of others through our willingness to sacrifice on their behalf.

Above all what ought to be the impetus behind this activity is a vision of love and selfless participation recognizing that God's care extends to all people, despite their condition in life, and that God will suffer in solidarity with them. Therefore the priorities of the church today must consider the good of all humans, and not only those who are good and just, "not only with one's neighbour, the member of one's own clan or party, not only with Christians, not only with saints, not only with the just" but with all who struggle.[16] Berry purports that this can only be done by reinventing the human at the species level through reimagining relationships with Earth. This will eschew anthropocentric hierarchies and instead prioritize planetary healing, affirming the creative tension that leads to life. "The historical mission of our time is to reinvent the human—at the species level, with critical reflection, within the community of life-systems, in a time-developmental context, by means of a shared dream experience."[17] To achieve this reinvention, the human must rely on our ability to learn new skills taught by the wider community, reflect upon and implement them to secure the ongoing existence of the Earth community.

Existing cultural expectations have burdened attempts at reinvention by espousing the belief that the planet is an object for use at human discretion, whatever the cost. Therefore a reimagination of relationships at the species level is necessary for recovering our innate awareness of the interdependence we share with other species.[18] This demands that we cooperate with others to bring this vision of the gospel to fruition, even if those others do not espouse our ideology or Christian ideal. Though we are bearers of our tradition, we share with all others of good will a concern for other people and for the intrinsic value of the human. With their help,

15. Schillebeeckx, *Church*, 184.

16. Schillebeeckx, *Christ*, 651.

17. Berry, *The Great Work*, 159.

18. Choi, "The Sacred Journey of the Earth Community," 81.

the gospel directives become refreshed, take deeper root in Christian life, and our hope in a better future seems more possible and believable. From this wisdom, new life is breathed into the sometimes-stagnant tradition and enables us to search for the new frontiers, already existing, in which we can make Christ's mission come alive.

In the despair of suffering, hope for a better and transformed future becomes real and believable only when it is united to the reality of poverty, wealth, and systemic oppression. In solidarity with victims and survivors who struggle to make meaning of their lives in the face of hardship, we bring the gospel to those in need. Having revitalized the gospel message through our actions, we create a "praxis of liberation" that prioritizes the common good by freeing those in bondage and enhancing the lives of those oppressed.[19] Essential to this praxis is a restructuring of cultural identities that reorients the human within Earth's context, exhibiting an awareness of the planet as primary and the human as secondary. Such reimagination must retrieve a sense of intimacy with Earth so that a deeper relationship is forged between humans and Earth and an awareness of the intricacies of Earth processes guides human behavior. An intelligible universe is present in every layer of created life and runs deep into the heart of the world. In the space of the universe, this intelligibility keeps all things knit together for the good of the whole. Central to this is a move away from a species hierarchy that sets the human over the rest of creation, disregarding the significance of creatures for their own subjective identity and the necessity of them for sustaining human life. Acknowledging this reality necessitates a shift from regarding only humans as subjects and the other-than-human as objects for mere instrumentality. Through our chosen lifestyles to gain power and feed our overconsumption, we have divided the planet into two realms: that of the human and that of everything else, as though we inhabit two separate realities that have no effect on each other.[20] Paramount to mending this deficient relationship is recovering a sense of the sacred in the universe penetrating every level of life. A reinvention of the human and a reimagination of cultural identities must start with a recovery with the numinous quality of the universe and, as a result, develop a way of life that honors this sacred reality through mutually enhancing relationships throughout all creation.

Seeking justice and peace, in addition to material needs, the church's mission requires cooperation from all people to combat this global tension with Earth. Such a comprehensive context where solidarity and liberation are sought bears witness to the gospel mandate to care for the poor while

19. Schillebeeckx, *Church*, 185.

20. Berry, "Christianity's Role in the Earth Project," 131, 134.

following the example of Christ, who is the model for us of God's compassionate nearness and will to save. God, who identifies with the suffering of humans and the injustices of history through Jesus' weakness and vulnerability on the cross, remains at work throughout history and offers grace through human action.

Ecotheological Anthropology

My experience with our farm family can be understood through the approach of Schillebeeckx, expanded to include the Earth community as seen in the work of Berry, who desires to reinvent and reimagine the human at the species level while prioritizing a sense of community that demands a certain lifestyle for ensuring its existence. Schillebeeckx underscores the twofold dimension of the resurrection that continues to inform the lives of Christians today as assembling a historical community gathered around a common, sacred task of renewing the world in an effort to have a better future. Read through the lens of the fifth Gospel, i.e., our personal narratives that enrich the Christian tradition, the historical community looking toward the far future becomes a broader Earth community spanning all time and seeking a better future for the entire communion of subjects.

In an ecological context, the importance of this compound insight takes on new meaning. While our relationships are intertwined and our existence deeply embedded in each other, the community tasked with the mission of working toward a better future for all planetary life is specifically human. The human is the only species aware of the possibility of its own extinction because of its complex consciousness and mode of understanding. Likewise, we are the species that is single-handedly dominating and destroying the planet through the overconsumption of natural resources and the gross production of harmful, irreparable waste. As a historical species developing from evolutionary emergence, humans in the context of Earth are tasked with the duty of intervening and halting this misuse of materials for the sake of the entire planetary community, including all other species who are powerless to protect themselves against our sophistication. Because of our adaptability, humans are capable of living differently and have historically been in much greater communion with the rest of the natural world. What is needed is a reinternalization of the rhythm of the universe for greater awareness of the Earth community so that we can achieve this hopeful future together.

The farm helps to illustrate the communion we seek with the other-than-human world and the ambiguity implicit in achieving it. Recognizing myself as an integral but not essential member of our ecosystem requires that I live in such a way that only enhances the life and good of the whole, encouraging the creative tension of the more-than-human family, even if that means placing restrictions on myself. While I need the land to be profitable, the land needs me to restrict my consumption of its resources. The restrictions that limit me permit the flourishing of life at a level beyond what is proper to the human and, ultimately, enhances my life. I am united to this place at both a bodily and affective level; I lose part of my identity when the health and vitality of our farm is compromised, which provides me with incentive for making such choices. Bound up together in the matrix of life in this place, the loss of individual creatures and species has an effect on me that is much farther reaching than I am aware and the absence is felt throughout the landscape.

Originally, our farm was part of the stretching oak savannas of the Midwest that provided a transitional ecosystem between the plains of the west and the forests of the east. Having been clear-cut to make room for fields and pastures, there are now few trees scattered throughout the countryside. The loss of this history and intuitive consciousness that grew deliberately in this space is palpable and makes me keenly aware that our contemporary existence is greatly divorced from that of our past. Even though some violence and destruction has been necessary for the promotion of agricultural industries, the devastation inflicted causes new problems, such as regular flooding in low-lying areas from a lack of trees. Nevertheless, liberation is still possible if humans unite themselves to Earth's cause and work for a sustainable future as one communion of subjects. Farming serves as a paradigm of the tension between Earth and the industrialized world where there is certain ecological cost for producing goods and gaining profit. Including Schillebeeckx's resurrection insights about forming a community for growing a better future allows humans to continue caring for our own needs while also caring for Earth's needs, as a healthy planet ensures the longevity of our own species' existence.

As Schillebeeckx highlights the importance of the disciples' mission of responding in solidarity to the needs of others, we live the mission of working toward a renewed ecological future by acting on behalf of those vulnerable to our destruction. Rooted in hope, we pursue the common good to enhance life on the planet for every member, as we love without distinction. To do this means, first and foremost, a renewed understanding of Earth as our primary teacher who directs humans toward more appropriate, sustainable living on the planet and to retrieve a sense of viability not dependent on our industrial ability to produce and consume.[21] On the farm the land shows us how it wants to be treated because the planet has existed for billions of years prior to human emergence. Therefore being in solidarity with Earth means intervening through action to protect creatures but also a willingness to deliberately abstain from action, instead allowing Earth to lead. In the reality of ecological decline, renewal of the world necessitates recognition of our total and utter dependence on the planet and attentiveness to the consciousness of other species, which dictates the specifications for their survival.

Finding ways to navigate between action and abstention while keeping Earth's needs central necessitates a human lifestyle committed to a better future. In an ecological context, the lifestyle of the kingdom seeks a more hallowed world for the sake of all creatures, and this living tradition of harmonious relationships with Earth must be recalled in the human heart and mind for responding to Earth's needs in creative and dynamic ways. This requires caring for specific species the way they ought to be cared for, even if that is counter to human proclivities. Above all, however, is the call for humans to sacrifice in our daily lives for the sake of ensuring the planet's ongoing existence by consuming less, living more sustainably, and using natural resources more prudently. On behalf of the communion of subjects, we are obligated to limit our indulgent lifestyles and instead suffer lack in solidarity with the wider Earth community, restricting our misuse and maltreatment of other life forms, so that a better future is realized for Earth right now that does not wait and depend on anyone but us. In doing so we assume personal responsibility for the health and vitality of other creatures and ecosystems, which ultimately enhances our own lives.

With Schillebeeckx, Berry helps interpret my experience of farm life by identifying how the lifestyle of the kingdom cooperates with the needs of the planet for devising a sustainable human species living in right relationship with Earth. Protecting natural resources from becoming commodities and recognizing their inherent rights allows Earth to become the authority that instructs relationships among species and sensitizes humans to the nuances of ecological changes, becoming the context for ongoing human formation to act on behalf of the planet. Just as the goal for the disciples was participation with Christ, the goal of humans today is cooperation with Earth, instead of control. On our farm in our more-than-human family, my home is more than a house with a human family; it is the totality of the other-than-human context: bald eagles flying, crops growing, and cattle grazing. Cooperating with these reorients my sensibilities to care about

21. Swimme and Berry, *The Universe Story*, 255.

their flourishing, thereby expanding my vision beyond what is good simply for me and instead considering human viability in the context of planetary life both now and in the future.

Conclusion

Schillebeeckx's commitment to a lifestyle of the kingdom focused on the praxis of solidarity and liberation with those who struggle, including the natural world, is how we might reinvent the human to promote a lifestyle of sustainability and cooperation with the planet. Looking ahead to a better future of renewed relationships between Earth and humans, we as a species that has emerged out of evolutionary history are compelled to participate with each other for the sake of developing connections with the planet that have previously been neglected. Reflection on our values and priorities ought to inform this process so that we engage only in life-enhancing relationships that prioritize the needs of all species.

The recovery of our human intimacy with the rest of creation is paramount to this lifestyle and a retrieval of our emergent history with the land is necessary for evoking the feelings necessary for reinvention. Like my relationship with our farm, being intertwined with natural space in the twenty-first century is complicated where industry and wealth are pursued with a singular focus. Reorienting ourselves so that we regard Earth as primary will be a difficult task requiring concerted effort but ultimately will contribute immeasurably to human flourishing and the good of the whole. Recognizing ourselves as part of the natural world is a critical first step in enacting this movement toward greater sustainable living on Earth as we anticipate our way into the future.

The complexities of domesticated land existing with wildness make this a unique relationship to navigate, as the control necessary for domesticated land necessarily requires human imposition. The farm exists because my ancestors cut down the trees and made it that way and we continue to maintain it. As a member of the Earth community, it is my job to ensure that relationships between the wild landscape, our farm, and us only seek to enhance each other for our ongoing survival of species. Bound up with each other, it becomes clear that we cannot live in isolation from the land, even if we tried, because it is so essential to our very existence and our identity as lovers of it. Finding my history in this space means that I am united to it at a deep and abiding level so that, more than the space in which I live, my farm becomes the space in which I recognize myself in my most fundamental state, that is, as a creature belonging to Earth.

Bibliography

Berry, Thomas. "Christianity's Role in the Earth Project." In *Christianity and Ecology*, edited by Dieter T. Hessel and Rosemary Radford Ruether, 127–34. Cambridge, MA: Harvard University Press, 2000.

_______. *Dream of the Earth.* San Francisco: Sierra Club, 1988.

_______. *The Great Work: Our Way into the Future.* New York: Three Rivers, 1999.

Choi, Kwang Sun. "The Sacred Journey of the Earth Community: Towards a Functional and Ecological Spirituality via the Cosmologies of Thomas Berry and Zhou Dunyi." Thesis: University of St. Michael's College, Toronto, ON, 2012.

Jenkins, Willis. *Ecologies of Grace: Environmental Ethics and Christian Theology.* Oxford: Oxford University Press, 2008.

Phillips, Peter. "Seeing with Eyes of Faith: Schillebeeckx and the Resurrection of Jesus." *New Blackfriars* 79.927 (1998) 241–50.

Schillebeeckx, Edward. *Christ: The Experience of Jesus as Lord.* Translated by John Bowden. New York: Crossroad, 1983.

_______. *Church: The Human Story of God.* Translated by John Bowden. New York: Crossroad, 1990.

_______. *God Among Us: The Gospel Proclaimed.* New York: Crossroad, 1983.

_______. "Naar een 'definitieve toekomst': belofte en menselijke bemiddeling." In *Toekomst van de religie—Religie van de toekomst?*, 37–55. Brussels, BE: Desclée de Brouwer, 1972.

Schreiter, Robert J. "Contrast Experiences." In *The Schillebeeckx Reader*, edited and translated by Robert J. Schreiter, 54–56. New York: Crossroad, 1987.

Simon, Derek J. "Salvation and Liberation in the Practical-Critical Soteriology of Schillebeeckx." *Theological Studies* 63.3 (September 2002) 494–520.

Swimme, Brian, and Thomas Berry. *The Universe Story: From the Primordial Flaring Forth to the Ecozoic Era, A Celebration of the Unfolding of the Cosmos.* New York: Harper Collins, 1992.

WILDERNESS AND WILD

Doing Theology with Snakes

Face-to-Face with the Wholly Other

—KIMBERLY CARFORE—

It was a cool, brisk summer morning in the foothills of the Blue Ridge Mountains of western North Carolina.[1] I was leading a girl's adolescent group through the wooded terrain of the Pisgah National Forest, nearing the end of my fourth year working as a field instructor for an at-risk youth wilderness therapy program. Students attending the program struggled with problems ranging from depression, anger/defiance, anxiety disorders, trauma, low self-esteem, emotion regulation difficulties, a history of sexual abuse, substance abuse, attention-deficit/hyperactivity disorder, and minor autistic spectrum disorder.

Six girls were asleep in their bags when I awoke to welcome a new day. Each day I woke up an hour and a half earlier than the students to enjoy some solitude before attending to the unique responsibilities the day presented. As the wildness of the land combined with the wildness of each student's psyche, it was impossible to predict what I might encounter. It was my job to maintain a high level of vigilance to keep the girls safe twenty-four hours a day, seven days straight, for two weeks at a time.

My morning routine consisted of a few stretches and a visit to the stream to wash my face and collect water for tea. After starting a fire with my

1. My intention in writing this piece is to recall a real-life event. Following Timothy Morton's "eco-mimesis," I chose to write in the past tense to maintain a certain level of authenticity. According to Morton, the more I attempt to evoke the atmosphere—where I was in the moment of the event—the more figures of speech I must employ, narrowing the rift between fiction and nonfiction. In other words, "my attempt to break the spell of language results in a further involvement in that very spell." Morton, *Ecology Without Nature*, 29–32.

bow-drill set, or assembling the portable stove (depending on what season it was), I would pour the stream water into my enamel camping cup, bringing it to a boil. While my black tea steeped, I did a few push-ups and crunches to get the blood flowing through my body. This morning was particularly chilly even though we were just coming out of the midsummer season. The thick, spruce-fir tree canopy above and dense, moss-covered understory of the Appalachian temperate rainforest created a consistently cool and damp climate.

In response to the chill, I put on my thick fleece hoodie and quickly slipped on my sandals, grabbing my metal cup on my way to the banks of the stream. After sliding on a patch of mud at the top of a steep grade, I found a shallow slope with dry footholds to safely descend toward the stream bed. I looked around to find a flat, sturdy rock to support my weight near a deep part of the stream. This perfect combination was sometimes hard to find and, depending on the campsite, impossible. I lucked out this time. Tapping the flat rock with my toe to ensure it wouldn't give when shifting my entire body weight onto the rock, I squatted down, in one smooth motion. As I squatted, I simultaneously took off my glasses with my left hand, placing them on a soft patch of dirt within arm's reach and began splashing water onto my face.

I gasped as the cold water touched my face. It was always so refreshing and I looked forward to this part of the day, every day. No matter how difficult the therapeutic and logistical dynamics of wilderness therapy could get I always felt fortunate having access to fresh, clean stream water. To be outdoors and touch wildness twenty-four hours a day seemed to make the stress of the work worthwhile. At the stream I let my mind wander wherever it wanted to go, letting the rush of the water guide it from thought to thought. I would daydream about my friends and family back home. I would think about office jobs and how strange that seemed to me at the time. I'd think about the person I was and wanted to become. I thought about the universe and how its dynamics are at work in social systems and structures, sometimes becoming blocked, repressed, or oppressed.

The outdoors provided the space away from society for students and instructors alike to reflect on themselves and their role in the world. Moments outdoors, in the wild, outside of comfort zones, revealed secrets of one's self and psyche within the backdrop of the wild. The inner landscape became the outer world where students could work on developing more positive coping strategies and behavioral habits that would help them when they returned to their everyday lives. The wilderness setting becomes a catalyst for personal growth and development. It became a catalyst for my work as a theologian.

Wild Slithers

Although I wished to sit alone near the stream for hours, it was time to continue my day and go wake the girls. Still squatting and still without my glasses on my face, I turned to my left to pick them up. Something felt off. My body lit up and I immediately froze. Even though I couldn't see it clearly, I knew something was there staring back at me. I felt it. Something registered in my blurred field of vision; something was not right. It is important to note that I am quite nearsighted. In fact, I purchase special eyeglass lenses in order to thin my prescription to keep the frail skin on my nose healthy under those would-be Coke bottle lenses. Without my glasses the rush of the stream and the world around appeared as a blur. Adrenaline pumped through my veins. In my bleary-eyed state, my eyes eventually focused and I registered the being I was encountering: a snake sat coiled directly in front of me.

Inches from my face, it did not move. I did not move. My left hand suspended in the air on its way to retrieve my glasses. Somehow, I placed them within centimeters of this snake's coiled body on my way to the stream. In that frozen moment, which could not have been more than a few seconds, this snake lapped its tiny red tongue at me. As gently as possible, I inched my left hand next to its body, exuding unobtrusive intentions. After picking up my glasses and placing them on my face, I could see that it was a venomous Copperhead snake.

This Copperhead is native to the Pisgah National Forest and is one of two venomous snakes in the area (the other is the Timber Rattlesnake). These reptiles are masters of camouflage and can be easy to miss. The irregular tan and dark-brown diamond banding on their back blends in easily with forest floor foliage. Accidentally stepping on them is a common story among those who have been bit. I was stunned having an uncommon and all too intimate face-to-face encounter with this snake.[2] One wrong move could prove to be deadly. At the very least it would send me straight to the

2. In this essay I am both playing off and challenging Emmanual Levinas's interview "The Paradox of Morality" where he is questioned on whether or not animals are included in his face-to-face ethics. Levinas et al., "The Paradox of Morality."

hospital. Considering that our group was many miles away from base camp, it could take hours before I could get to a hospital.

While I immediately intuited its danger when my glasses were off, seeing this snake opened me to feelings of profound dread. Without breaking eye contact (because that was my instinctual response) and still in the squatted position, I placed my hands in the air in surrender, but to this day I cannot recall if this happened in the phenomenal world or only in my imagination. Somehow this posture of surrender was in my field of intention. It *was* my field of intention. It was almost all I had as a defense mechanism. The air felt thick, and every possibility seemed heavy with consequence. It felt like there was no myself, and there was no snake; there was simply this moment; this event; the "thereness" of it all. There was no decider, only decision; there was no thinking, only thought. There was no space for thought to wander, to think, to swing back and forth as the grasping monkey mind, nor was there space to decide. A decision was made within a space-time that was merged. Space and time in this moment was more continuous than in ordinary life. Decision was closer to instinct, but even this word does not relay the truth of this experience. I could say it was a corporeal, embodied, or affective response, but the experience resists easy categorization.[3]

As this snake sat inches from my face and I looked into its tiny black eyes, I experienced (what I later deduced as) an ethical imperative. Speaking different languages—my verbal, cognitive-centric, rational language juxtaposed with this snake's nonverbal wildness—there was no lack of communication. We both communicated the same imperative, "Don't kill me," sounding like the divine commandment, "thou shall not kill," also translated as "you shall not murder" (Exod 20:13).[4] Within this space, and exuding a mutual ethical imperative, I backed up. The snake did not move an inch. It did not uncoil. It did not slither closer to me. It did not follow me. It simply sat there, being itself.

3. I agree with John Llewelyn's interpretation of Levinas's "face to face" in that the face-to-face encounter "cannot be named or nominalized. It cannot be said." I expand on his statement that the "proximity of the face to face . . . is not a topic of theology." Llewelyn, "Levinas, Derrida, and Others Vis-à-vis," 147.

4. In his work, Levinas recalls often that the face of the other always presents itself as a commandment, "you shall not murder." Derrida draws attention to the important distinction between murder and killing—murder is something humans do to other humans, namely homicide. An animal is not murdered, it is killed. Derrida critiques Levinas's humanistic ethics, "Levinas insists on the originary . . . character of ethics as human." Levinas's ethics places the human before animals and "never looks at him to say 'Thou shalt not kill.'" Derrida, *The Animal That Therefore I Am*, 110, 108.

Just by being itself this snake won its territory. I had nothing but respect and fear for this animal. This reptile exuded a power I've never experienced before. These species are much older than we are. Early ancestors of snakes appeared over 300 million years ago. They have survived three mass extinctions. It has seen more than I have. Its skin had a tightness that displayed perfection. Each scale seemed to tell a story, and there were thousands on this snake's body. Evolution got this design right and stuck with it, perfecting the spaces between each scale over hundreds of millions of years.

While this snake won its territory, due to logistical concerns we were forced to spend the day and another night at this campsite. When night fell and we prepared ourselves for bed, off in the distance a student screamed that something had brushed up against her foot. I immediately met eyes with my co-instructor and with fear and mutual understanding we made our next move. Sure enough, as I scanned the ground for any disturbance in my field of vision, I saw leaves rustle and something moving through them. There it was, slithering on the forest ground. It moved slowly, nonviolently, making small s-patterns about thirteen feet from us, paralleling our path. My stomach dropped. I turned to this snake, focusing all my thought and intention to it, and spoke directly: "Please spare yourself, friend. This is your warning. I do *not* want to kill you. But I will, if I have to, if you get too close."

Ten minutes passed. I approached the branch that established the boundary between the wild oak forage of the forest and the groomed, soil-exposed ground of our campsite. The snake had just rounded the corner at the end of the branch and began darting towards us in tight s-, almost z-patterns quite aggressively. It felt like this snake was teasing me with its movements. It was making fun of my inferiority. The time for words had passed; it was now time for action.

A blur of adrenaline, instinct, and action ensued. I do not remember exactly what happened, but my body was on fire. As I held this snake's neck to the ground with my walking stick, my co-instructor, in one smooth, quick movement, decapitated this snake with a shovel. Separated from its body, the jaw snapped a few more times—its final attempt to fight back. Energy pulsed through its long corpse, causing it to squirm in random positions, back and forth, until it coiled into its final resting position, which happened to be in the shape of an infinity symbol. I felt the weight of certainty as I knew at that moment that infinity symbol signaled the end of my career as a wilderness therapy field instructor. Although I had no training in reading symbols, I intuited what this meant for me. I ended my career knowing I was "uniquely responsible" for this snake's life, and perhaps more profoundly, the event of this snake's death.[5]

Mourning the creature, and confused about the event, I sat hunched over this snake and cried. It felt like I was vomiting tears of disgust; disgust with myself as I had always considered myself a giver, not a taker. This was my ethic in life, the rule I lived by, to always leave things better off than I received them. If someone or something needed something, I gave. If there was a cause, I gave. So much so that this had become an identity of mine; a category so solidified that I experienced shock the moment it shattered. "I am a taker," I thought. I could only feel deep sorrow letting these words sink into my heart. But when such words began to sink in, the tears stopped, caught within a net of paradox. The moment these words entered my being became an overwhelming moment filled with pain, sickness, truth, knowing, rejection, and acceptance. Rejection-acceptance-rejection was the dance of emotion. I was shocked. The ambiguity of the situation was nauseating. I wanted to vomit to experience release but I was empty; emptied of any sort of certainty about almost anything. There was nothing solid to hold onto and spit out, except for my tears.

I felt furious about being forced to make what seemed to be an impossible decision. I felt this decision had been unfairly presented to me—I did not ask for it; I did not want it. Was it the death of my identity, my ethical status as a "giver" or the ending of this snake's life that was so impossible to digest?[6] I could not digest the feeling of being a taker—to take another creature's life with no intention of eating it, or using it for any other purpose other than my selfish human needs. If I could use this corpse, as it represented the telos of this snake's entire life, if I could bear witness to its death, its meaningless suffering, then maybe I could recycle the energy of everything this snake represented back into the cycle of Earth. Life to death

5. Llewelyn, "Levinas, Derrida, and Others," 148. Llewelyn draws from Exodus 33:11, "the Lord spake unto Moses face to face." When the Other commands, I am "ethically and religiously bound to answer . . . I am uniquely responsible." For more on the concept of "the event" see, Caputo, *The Weakness of God*. According to Caputo, the event exposes God as weak, uncertain and unstable, and "a trace of a voice." Ibid., 97.

6. The death of this snake took place in spite of its commandment towards life, "don't kill me." I imagine if this animal remained a faceless other, if I wouldn't have encountered this snake hours earlier, it wouldn't have been such an ethically traumatic event. In other words, recognizing my kinship with this snake, experiencing the wholly other in this other, made the event of its death utterly confusing and devastating. The ethical tension resulting from the double exposure of vulnerabilities resulted in my experience of indigestion. While Derridean and Levinasian ethics perhaps assume that peace is a possibility, my encounter with this wild Earth other demonstrates something a bit different, something closer to an ethics of ambiguity (a term borrowed from Simone de Beauvoir).

could be rebirthed into life-death-life. If I could simply *do* something, maybe the overwhelming ambiguity I experienced might become meaningful. If I could make sense of this nonsense the situation might become bearable. If I could take action, or control, then maybe the pain would go away.

What did I *do* then? I buried its body. I could not eat it because I could not start a fire that late into the night. I could not leave it for an animal to eat because it would attract human predators (bears, badgers, coyotes) into our camp. Buried under the ground no animal would find it. It could not be "utilized" back into the food chain. It was a trace signifying an event passed.[7] The decision had been made. Its life had been taken by me. There was no turning back. I felt weak. I felt violent. I felt both that I had saved and failed, helped and harmed my group.

We first buried the head to prevent venom poisoning. Decapitated snakes are known for retaining reflexes and will continue to bite, releasing the venom left in their glands. I uncoiled this snake's beheaded carcass and caressed it, speaking to it, mumbling prayers. I turned it on its black-diamond back revealing its soft, creamy white underbelly, where I skinned it with my bare hands and a dull Shrade Old Timer knife. This was my way of respecting, remembering, and honoring this snake. It felt like all I could do to render my emotions allowing the shock to subside. After I cut a vertical slit through its neck I pulled the skin back, peeling the rest of the skin off this snake's flesh.

We buried the body and said a few prayers. I wandered down to the stream, and washed the skin in the water where I first encountered this snake. I remained awake for most of the night, maintaining vigilance for I don't know what—for anything that might come: an insight, a bear, a meaning. The skin remains mounted on a stick above the entrance to my apartment, at the threshold of my home: the liminal space between here and there, home and foreign, domestic and wild, self and other. I tattooed my arm with a block infinity symbol in solidarity with my decision to kill that snake, and to never forget this one snake's life, nor this remarkable event. In a sense, tattooing my skin was marking the remarkable, or marking that which could not be marked. An experience that moves beyond all categorization cannot actually be marked. There is something infinite about experiences that cannot be catalogued or captured by understanding. The otherness I encountered in the face of this snake shook me to my core, and it still haunts me today, shaping my interpretation of myself, snakes, and every

7. "The trace" is a Levinasian concept that Derrida exposes in deconstruction, undoing the metaphysics of presence. Undoing the presence/absence dualism, the trace signifies a rupture in presence, which is always haunted and complicated by non-presence.

other, including that which is wholly Other—God. When I do theology, I do it with snakes.

Wild God Talk

What stands out in my particular experience is the mark of infinity that was traced on an experience of my own finitude. If I am to begin interpreting my encounter with this snake, the point would *not* be to interpret an encounter with the infinite, or to otherwise bring infinity into finitude, dark into light, or unconscious into consciousness.[8] This would be assimilating the wholly otherness of the infinite into one's own horizon. Assimilating the other into the self does violence to the other, closing off the possibility of justice. However, there is no non-mode of interpretation that does not already assimilate an event into one's own horizon of experience. This paradox of otherness is one I came to articulate through Jacques Derrida's deconstruction.

As Derrida affirms, "There is not narcissism and non-narcissism; there are narcissisms that are more or less comprehensive, generous, open, extended."[9] In other words, you can't jump over your own shadow to step into the world of the other. Derrida's philosophy, specifically his method of interpretation and literary analysis known as *deconstruction*, offers a theologically sound mode of interpretation leaving open the possibility of a justice to come. While Derrida does not necessarily *do* theology, his sense of the deconstruction of religion offers a postsecular *religion without religion* that opens up the name of God to the names of multiple others, indeed, every other.[10] If the work of deconstruction opens structures to be more "comprehensive, generous, open, extended," then a deconstructive *religion without religion* would be the practice of extending religiosity to welcome more others, including snakes.[11]

8. The point, if I were to need to articulate a *point*, would be towards justice.

9. Derrida, "There is No *One* Narcissism," 199.

10. While Derrida himself was not necessarily a theologian per se, scholars of Derrida—especially John D. Caputo—have interpreted Derrida's work as religious and theological. See Caputo, *The Prayers and Tears of Jacques Derrida*. According to Caputo, Derrida "practices a secret religion." Caputo, "Before Creation," 91. In addition, Edith Wyschogrod and Hélène Cixous have offered the term "postmodern saint" to describe the life and work of Derrida. For more on this see Cixous, *Portrait of Jacques Derrida as a Young Jewish Saint*; Wyschogrod, *Saints and Postmodernism*.

11. "Religion without religion" is a term Caputo uses to describe Derrida's religion. According to Caputo, Derrida "has a religion" but does not follow specific religious rituals. He "speaks of God all the time" but does not speak to "religion's God." Caputo,

Deconstruction, as developed by Derrida, is not destructive. It is about doing justice. It opens to events of justice by welcoming the arrival of every other. In this sense, deconstruction activates a religious structure—a messianic call for justice—while holding in suspense any commitment to the biblical context that gave rise to that structure. In other words, Derrida's deconstructive sense of justice seeks to "remove a biblical surface from a messianic structure."[12] "*Deconstruction is justice*" is a phrase that is religious and, at the same time, without religion.[13]

Recovering a messianic structure without a determinate messiah or messianism might do justice to religion in a postmodern, postindustrial, globalized world. As some religions might be awaiting the arrival of the second coming of Jesus Christ as the messiah in order to mark the arrival of the wholly other, individuals practicing their religion might miss the arrival the messiah, or of multiple messiahs throughout their lived experiences. Practicing a *religion without religion* would be practicing messianic justice without a determinate messiah or messianism. In other words, the arrival of justice would not necessarily be the second coming of Jesus Christ. It does not rule out the possibility of Christ as the messiah, but other others might slither in as well.

Opening to otherness ("alterity" in deconstructive parlance) marks the arrival of an event of justice. Interpretation closes oneself within neatly marked categories—infinite/finite, sacred/profane, human/animal, religious/secular, self/other. Assimilation then easily becomes a mode of violence—marking the self from the other opens onto a slew of misinterpretations that can all too easily fall into xenophobia, racism, sexism, and speciesism. Ambiguity, uncanniness, paradox—these are modes of postsecular *religion without religion* opening towards justice. Welcoming the uncanny, ambiguity, and paradox marks the arrival of an event of justice. Deconstruction offers a method that does not close off violating interpretations of the other, but rather leaves open the paradoxical arrival of otherness. The nausea I experienced encountering this Copperhead snake in the Pisgah National Forest signifies how overwhelming ambiguity is. Letting the ambiguous be without making it something it was not—something bearable for me—was an unbearable experience that somehow I bore.

Prayers and Tears, xvii. The purpose of a *religion without religion* would be to welcome an event of the impossible. For example, welcoming a saint is not necessarily impossible since it has already occurred. Welcoming a snake as saint, or messiah, would be closer to something like welcoming the impossible.

12. Ibid., 135.

13. Derrida, *Acts of Religion*, 243.

My encounter with the Copperhead snake was an encounter with the *arrivant*—a French word which can be translated as "comer," "newcomer," or "one who arrives." Derrida considers the *arrivant* as the arrival of the wholly other—the "absolute and unpredictable singularity" that interrupts relations that humans have with one another, as well as between humans and Earth others.[14] The wholly other, according to Derrida, is palindromic, "*tout autre est tout autre*," which translates reversibly as "every other is altogether other" and "altogether other is every other."[15] This demonstrates the paradoxical singularity and universality of God as wholly other, "Every other (one) is God," and "God is every (bit) other."[16] *Tout autre est tout autre* signifies that "every other is singular" as well as "every one is each one."[17] "Every one is each one" implies a universality, whereas God "is to be found everywhere," specifically where "there is a trace of the wholly other."[18]

An ethics of alterity marks an opening where the other overflows proper boundaries of self/other. As I experienced in the wild, alterity is not reserved for humans, but applies equally to all life on Earth and, indeed, to all beings. An encounter with alterity demands respect. My encounter with this Copperhead snake overflowed these neatly human concepts of reverence and respect. The ethical call emanated from the wildness of being which does not settle neatly into categorization. "Don't kill me" was my plea to the snake, as well as the plea I experienced emanating from the snake. In the event of our encounter, I did not choose to think the thought "you shall not murder." Rather, the thought erupted from some place, someone, else. You could call it God; you could call it the ethics of the wild. This call emanating from my encounter with this snake was a call for a response and responsibility for every other as wholly other. Hearing that call, how could I not find myself deconstructing conceptions of God, opening to a non-anthropocentric theology? The encounter between myself and this Copperhead provided an experiential invitation into deconstructive theology. This encounter broke open my concept of God as a transcendent, monotheistic God to new possibilities accounting for the divine otherness (alterity) overflowing this event. I could not wrap my head around this event. This encounter broke through concepts of God that I had lived and acquired growing up in the Catholic Church. My experience simply did not fit. It had

14. Derrida, *Specters of Marx*, 28.

15. Ibid., 195, n. 37.

16. Derrida, *The Gift of Death*, 92, 87.

17. Ibid., 87.

18. Ibid., 78.

a force, perhaps the weak force of God, which forced itself out of preconceived notions of God.

Facing this snake broke open my concept of God in two ways: ontologically and ethically. After the event I considered God as the wholly other Being that is and interpenetrates all beings, not simply humans, and the ethically compelling force that runs through all beings. This ethically compelling force is the call of justice. This sense of justice accompanies the politics of a democracy to come. It is always to come, for if justice arrives it "rests on the good conscience of having done one's duty," and "loses the chance of the future."[19] This "to come" makes possible the space for the arrival of the wholly other, which is never finished arriving, for "[t]he *un*conditional is always to-come."[20]

It is overwhelming. Who am I when I feel responsible for the wholly Other issuing forth from my encounters with every single other? In the words of Derrida, "[w]hat is the 'I,' and what becomes of responsibility once the identity of the 'I' trembles *in secret*?"[21] To help demonstrate this question I turn to James Hatley, who describes his experience facing a bear attack. In this encounter, he states, "I am placed utterly outside of myself, to the point that *I am an other* and/or the other is so utterly inside me that *no space remains where I can be merely myself*."[22] However, it is not merely terror that accompanies face-to-face encounters with wild others. The "uncanny goodness" of being edible to wild animals was palpable to me. The same body that encompasses one's being is the body that could be inside another being's body within seconds, so that other body can endure in its life. In other words, as the *Taittiriya Upanishad* states, "From food are born all beings which, being born, grow by food. All beings feed upon food, and when they die, food feeds upon them."[23]

An experience of being food undoes tidy dualistic categories of self/other, subject/object, inside/outside, human/animal. Ecofeminist philosopher Val Plumwood demonstrates this in her famous essay, "Being Prey," which accounts for her near-death experience while being attacked by a crocodile. Plumwood states, "In the moment of truth, abstract knowledge becomes concrete."[24] In this "moment of truth" there is an "extreme heightening of consciousness evoked at the point of death" where "extraordinary

19. Derrida, *Specters of Marx*, 28.
20. Caputo, "Before Creation," 97.
21. Derrida, *The Gift of Death*, 92.
22. Hatley, "The Uncanny Goodness of Being Edible to Bears," 21.
23. Prabhavananda and Manchester, eds., *The Upanishads: Breath of the Eternal*, 55.
24. Plumwood, "Meeting the Predator," 10.

visions and insights" appear. It hits you that "you were completely wrong about it all—not only what your personal life meant, but about what life and death themselves actually mean."[25] The experience of being prey is the experience of being mortal. It is the experience of being shaken outside of oneself and of everyday conceptions of reality. It is revelatory. This revelation transgresses the boundaries of the human, crossing every other, from divinity to animality, a crossing Derrida names with the portmanteau "divinanimality."[26]

Conclusion

In conclusion, my encounter with the Copperhead snake was *not* an encounter with God as an anthropomorphized being in the sky (transcendental monotheism), but God as the ethically compelling trace that runs through all beings, insisting on justice. My life changed after this encounter; it took on a new direction. I was not the same person after the encounter. I had grown up following an idol-type figure—a temperamental Father who seemed a far cry from the insistent call of the wild wholly Other. It was not the really real, the other who arrives in and as every other, that keeps coming, always already demanding justice. That dubious Father figure has ceded his authority, and now I do theology with snakes and with people like Derrida. I should add that Emmanuel Levinas is not far from my thinking here. His sense of ethics and religion as an encounter with alterity exemplified in face-to-face relations is an important influence on Derrida and on my own theology.[27] However, Levinas remains too anthropocentric in his thinking. For him, not every single other bears the compelling trace of a face. Consider this remark. "I don't know if a snake has a face. I can't answer that question."[28] With Derrida, I *can* answer that question. A snake has a face; it harbors the radical alterity of the wholly Other, as does every being we encounter on Earth.[29]

25. Ibid., 11.

26. For more on this see Moore, ed., *Divinanimality*.

27. On religion as a relation with alterity, i.e., a "relation without relation," see Levinas, *Totality and Infinity*, 80.

28. Levinas et al., "The Paradox of Morality," 172. Levinas continues: "A more specific analysis is needed."

29. In Derrida's *The Animal That Therefore I Am*, 105–18, Derrida suggests that Levinas has all the resources to embrace the other as animal but does not. Derrida analyzes Levinas's humanism, inferring that "a more specific analysis is needed" to be "an admission of nonresponse," declining responsibility, 108. Furthermore, Derrida draws

I had thought the path to God was a path inside or beyond, but then I realized that God is in the face of a snake. The wholly Other shows up in every single other. I used to practice purifying my thoughts in order to attain spiritual heights. Now, I practice attending to the otherness of every other, engaging in wild God talk, and the ongoing work of justice. I'm still haunted by the wild. My vision is still blurry as I look out at a world in which every other is wholly other. Feeling uncertain and uncanny, I have no idea what my theology is, if it can even be said that I have or possess a theology. Without knowing, without having, without seeing, as Derrida says, "*sans savoir, sans avoir, sans voir*," whatever place theology has for me remains in the wild.

Bibliography

Caputo, John D. "Before Creation: Derrida's Memory of God." *Mosaic* 39.3 (September 2006) 91–102.

———. *The Prayers and Tears of Jacques Derrida: Religion without Religion*. Bloomington, IN: Indiana University Press, 1997.

———. *The Weakness of God: A Theology of the Event*. Bloomington, IN: Indiana University Press, 2006.

Cixous, Hélène. *Portrait of Jacques Derrida as a Young Jewish Saint*. Translated by Beverly Bie Brahic. New York: Columbia University Press, 2004.

Derrida, Jacques. *Acts of Religion*. Edited by Gil Anidjar. New York: Routledge, 2010.

———. *Specters of Marx: The State of the Debt, the Work of Mourning, and the New International*. Translated by Peggy Kamuf. New York: Routledge, 1994.

———. *The Animal That Therefore I Am*. Edited by Marie-Louise Mallet. Translated by David Wills. New York: Fordham University Press, 2008.

———. *The Gift of Death*. Translated by David Wills. Chicago: The University of Chicago Press, 1995.

———. "There is No *One* Narcissism." In *Points: Interviews, 1974–1994*, edited by Elisabeth Weber, translated by Peggy Kamuf, 196–215. Stanford, CA: Stanford University Press, 1995.

Hatley, James. "The Uncanny Goodness of Being Edible to Bears." In *Rethinking Nature: Essays in Environmental Philosophy*, edited by Bruce V. Foltz and Robert Frodeman, 13–31. Bloomington, IN: Indiana University Press, 2004.

Levinas, Emmanuel. *Totality and Infinity: An Essay on Exteriority*. Norwell, MA: Kluwer Academic, 1979.

Levinas, Emmanuel, Tamra Wright, Peter Hughes, and Alison Ainley. "The Paradox of Morality: An Interview with Emmanuel Levinas." Translated by A. Benjamin

attention to Levinas's choice of animal—the snake—a creature that carries "immense allegorical or mythical . . . biblical and poetic weight," making attributing a face to this creature "highly improbable," 110.

and T. Wright. In *The Provocation of Levinas: Rethinking the Other*, edited by R. Bernasconi and D. Wood, 168–80. London: Routledge, 1998.

Llewelyn, John. "Levinas, Derrida, and Others Vis-à-vis." In *The Provocation of Levinas: Rethinking the Other*, edited by R. Bernasconi and D. Wood, 136–55. London: Routledge, 1998.

Moore, Stephen D., ed. *Divinanimality: Animal Theory, Creaturely Theology*. New York: Fordham University Press, 2014.

Morton, Timothy. *Ecology Without Nature: Rethinking Environmental Aesthetics*. Cambridge, MA: Harvard University Press, 2007.

Plumwood, Val. "Meeting the Predator." In *The Eye of the Crocodile*, edited by Lorraine Shannon, 9–21. Canberra, Australia: Australian National University Press, 2012. A version of this originally published as Val Plumwood, "Being Prey," *Terra Nova* 1.3 (Summer 1996) 32–44.

Prabhavananda, and Frederick Manchester, eds. *The Upanishads: Breath of the Eternal*. New York: Mentor, 1957.

Wyschogrod, Edith. *Saints and Postmodernism: Revisioning Moral Philosophy*. Chicago: The University of Chicago Press, 1990.

Living, Local, Wild Waters

Into Baptismal Reality

—LISA E. DAHILL—

"Our collective future is teetering between drought and flood."
—Judy Natal

I set out for my first voyage late one June afternoon. . . . I was astonished to realize that at kayak level in the water, the surrounding trees blocked all urban view except for the occasional bridge overhead, and these trees were filled with warblers, crows, vultures, and hawks. The creek itself was populated with ducks and herons, as well as the occasional kingfisher. Around the roots of the trees that first magical dusk voyage I saw a skunk, a mink, a family of raccoons climbing one of the sycamores overarching the river. But it was in the water—which from street view above I had assumed was basically dead—that the miracle happened. The golden light of the late afternoon somehow hit at just the right angle that the stream's depth lit up, and I was stunned to see masses of tiny fish darting in union, shadowy carp, many mollusk shells, riparian plants, some bass, even a water-snake. The water was clear as light, rich with nymphs and organisms, each milliliter of this urban stream full of life, and I knew for the first time that the water supporting life is *itself alive*. Living water is wild water. And I sensed how urgently Christianity and Christians needed to be baptized fully *into these actual waters*, these living waters. Over the years I kept kayaking, and began

> wading and swimming as well, getting to know this creek. I began designing seminary rites of baptismal remembrance along its banks, and I learned how the herons and riparian creatures and homeless folks at the edges of our circle pull ritual language out into all sorts of new connections.[1]

That first trip flowed into countless more, as I paddled through the creeks, rivers, and lakes of central Ohio, in Lake Superior, and in wild Western rivers (the North Fork of the Flathead in Montana, the Green River west of Denver) and the Pacific. This literal immersion in living, local, wild water—beginning in a sabbatical year that included also a month in a North Cascades wilderness context in December, several other multi-day outdoor immersions, and the transforming experience of reading David Abram's *The Spell of the Sensuous*—opened me into the forms of thought emerging ever since.

It is from Abram's book that I draw the method of this essay. Regarding the intellectual process that gives rise to his writing, Abram notes:

> It is a way of thinking that strives for rigor without forfeiting our animal kinship with the world around us—an attempt to think in accordance with the senses, to ponder and reflect without severing our sensorial bond with the owls and the wind. It is a style of thinking, then, that associates *truth* not with static fact, but with a quality of relationship.[2]

What would it mean to do theology in this style: to think and speak as clearly and precisely as possible about the Christian tradition and its texts, beliefs, categories, rituals, from within experience of the natural world on its own terms, "without forfeiting our animal kinship with the world [and] in accordance with the senses"?

I learned to swim in the ocean and spent my childhood immersed in the surge and tug of the tides, bodysurfing and diving under waves, and at least once in high Pacific surf nearly drowning. At age nine I was baptized into the Christian church in a temperature-balanced indoor baptistry. Both categories of experience—ocean swimming and baptism—involved bodily immersion in water, but nothing in the larger sets of stories or meanings attached to each led me to see any necessary connection between them. Through later ordination and scholarly vocation, I held together love of the natural world and Christian faith, and over the years I embraced the

1. This text is taken from my essay "Into Local Waters." The oxygenation of those urban waters (Alum Creek in Columbus, Ohio) had received a major boost through the recent removal of most of the low-head dams impeding the creek's flow.

2. Abram, *The Spell of the Sensuous*, 264.

incarnational and sacramental interconnections between them. Only relatively recently, however—in this series of untamed water-encounters—did the faith into which my baptism immersed me dissolve for good into the larger fullness and wildness of the natural world. As I ponder the disconnect many experience between Christianity and the natural world, I have begun rethinking the practice at the threshold and center of Christian life: baptism itself. Into what reality does—and should—Christian baptism invite humans?

This essay is part of a larger project of "rewilding Christian spirituality" centered in the proposal to restore the early church's practice of baptismal immersion into living (i.e., flowing natural) water. Such practice baptizes persons

> not only into the human Body of Christ, but into bodies of real water with their own public political and ecological life, and into the Body of God in Sallie McFague's sense: the biosphere. . . . Practicing baptism in this way . . . is harder than baptism in a room. It's more dangerous, more public, more political, more euphoric; it binds participants to the literal water of a given place at the heart of their experience of Christ, and it binds community members to one another and to the human and non-human life of that place in an unforgettable intimacy. Such practice takes seriously the *biological and literal* dimensions of the Word becoming flesh, becoming matter, taking on and permeating and filling all created life.[3]

This practice implicates communities in the health of the water and watershed, recognizing that entrusting infants and adults to these waters requires ongoing collaboration with scientists monitoring a given watershed, activists safeguarding it, other humans living near these waters, and patterns of habitation, pollution, species migration, zoning, and flow affecting it all.[4]

One way I understand Abram's approach is in learning to encounter the Book of Nature as having revelatory authority equivalent to—or, when they collide, greater than—the Book of Scripture (and/or tradition) for Christians.[5] The language of the "Book of Nature" is ancient in Christianity,

3. Dahill, "Rewilding Christian Spirituality." See McFague, *The Body of God*. Sacramentally oriented traditions insist on the primacy of the literally embodied dimension of Word and sacraments, inseparable from their symbolic function; see, e.g., Lathrop, *Holy Ground*, and Gibler, *From the Beginning to Baptism*.

4. These questions also animate other recent work of mine. See Dahill, "Alive Together."

5. I have elsewhere defined "the Book of Nature as all forms of knowing (of God or reality) whose source is the natural world itself, both in direct sensory engagement

articulating the christological insight that the same Word active in Genesis 1 is the one enfleshed in Jesus Christ (John 1) and speaking through Scripture. I seek "to ponder and reflect without severing our sensorial bond with the owls and the wind," to do theology from within "our animal kinship with the world." In the Lutheran tradition I inhabit, such thinking based in the creek in addition to Scripture or tradition is still unusual. In a time of ecological emergency, however, in which voices from every tradition and species are calling us to "ecological conversion," such larger thinking from the Book of Nature is surely necessary.[6]

And this risk of taking seriously the Book of Nature as having a role theologically critical of insight based solely in Scripture and tradition is one that Lutherans ought to be able to embrace. Not only does Luther's own witness develop forms of reliance on the Word of God that challenged the interpretive hierarchies of his tradition.[7] But Dietrich Bonhoeffer provides an even more explicit basis for this new work. In his *Ethics*, Bonhoeffer names Christian dualism—by which Christians conceive of reality within two separate spheres, "God" and "world"—as a "Colossus obstructing our way."[8] He articulates a nondualistic vision of reality by which God and world are truly integrated in Christ: "There are not two realities, but *only one reality*. . . . [T]he whole reality of the world has already been drawn into and is held together in Christ. . . . [T]he world, the natural, the profane, and reason are . . . seen as included in God from the beginning."[9] This is a vision so con-

---

with that world and as refracted through scientific study of the world." In the same essay I contrast this with a definition of "the Book of Scripture as all forms of knowing (of God or reality) whose source is the Bible: both the direct study of the book itself and as that opens up into millennia of Christian texts, sermons, artwork, practices, community structures, and institutions: the Christian imagination itself, sourced in the Bible's revelation of divine presence through Jesus Christ in, with, and under all that is." Dahill, "Into Local Waters." For more on the history of this category, see Clingerman, "Reading the Book of Nature."

6. Francis, *Laudato Sí*, paras. 216–21.

7. The volume *Eco-Reformation* (ed. Dahill and Martin-Schramm) collects a range of essays by Lutherans exploring ecological contributions of this tradition, including many resources from Luther. The question of Luther's contested but generally suspicious stance toward "natural revelation" is one I hope to take up more explicitly in the future. For now, it's fair to say that I am using Luther's scriptural method to read him in turn (i.e., using Luther to critique Luther). That is, I am giving greater interpretive priority to Luther's incarnational and sacramental principle (*finitum capax infinitum*, by which he asserts that natural reality has the capacity to bear the fullness of divine presence and revelation) than to his insistence on *sola scriptura*. Such a move reflects Luther's own Augustinian roots; cf. Augustine, *De Genesi* (1.19.39), *ACW* 41:42–43."

8. Bonhoeffer, *Ethics*, 55.

9. Ibid., 58–59.

vincing to him that he is unable to grant fundamental separation between God and world even in the heart of Nazi Germany.[10]

Bonhoeffer did not live to develop what it might mean to think ecologically within this framework, and certainly he never articulates any privileging of nature over Scripture in the way this essay does. But as I have thought with his insistence on nondualism over many years, I am recognizing that it opens the door to the Christian thinkability of a hermeneutics by which insights from the Book of Nature are in fact allowed to critique, recast, or expand forms of thought and practice derived from Scripture and tradition.[11] If indeed reality is nondualistic, such that one cannot defend a supernatural "revelation" incompatible with biotic interrelationship—that is, if (in Christian terms) the same Word of God that speaks in Scripture also speaks in trustworthy ways in and strangely permeates all of natural reality—then surely direct experience of the natural world can have a theologically critical role in Christian experience.[12] And this role can, perhaps must, be one of revelatory priority, preceding Scripture. As Thomas Berry put it, "The universe, the solar system, and the planet Earth in themselves and in their evolutionary emergence constitute for the human community the primary revelation of that ultimate mystery whence all things emerge into being."[13] Here therefore, I wish to let my Alum Creek kayak experience open baptismal implications of immersion in *living* water, *local* water, *wild* water, including the riskiest theo-logos of all, namely pondering—still within the creek and its flow—the question of divine life into or by which we baptize. The test of this essay is not primarily in the strength or weakness of its various insights but in the degree to which its process generates new insight: does it work to think theologically more or less directly from the Book of Nature? What, if anything, makes such thinking Christian?

10. See especially 64–68 in his *Ethics*.

11. For work on Bonhoeffer as an ecological theologian, see Rasmussen, "Bonhoeffer as Ecological Theologian," and Martin-Schramm, "Bonhoeffer, the Church, and the Climate Question." My articles "Bio-Theoacoustics" and "The View from *Way* Below" use key aspects of Bonhoeffer's thinking (his critique of abstraction and this insistence on one reality, respectively) in thinking ecologically. The latter essay also shows how this insight builds on Bonhoeffer's work in *Creation and Fall* on the emergence of "Zwiespalt," or splitting, dualism, "two-ing," in the accounts of Genesis 3.

12. Of course, as with assertions of "revelation" in Scriptural reading, here too the heart of the question is hermeneutics: how exactly does one credibly discern, trust, or assert the content of such reading? On this, see the emerging fields of ecological hermeneutics: for philosophically based work on these questions, one might begin with *Interpreting Nature,* ed. Clingerman et al.; for biblical engagement, see Habel, "Introducing Ecological Hermeneutics," along with the Earth Bible Project and works inspired by or critical of this project.

13. Berry, "Twelve Principles," 216.

Water

Living Waters

The heart of what struck me in that first kayak journey into Alum Creek was the shock of discovering truly *living water* in a polluted urban creek. If this creek can contain so many species of all kinds, imagine how much must fill less degraded waterways—imagine, that is, how much biotic *life* of all kinds and phyla is implied in the term "living water." It is into this fullness of life that baptism is meant to immerse Christians, a truth largely overlooked in the practice of indoor baptism in small bowls using dead treated, sterilized water. The early church's baptismal practice famously specified immersion in living (i.e., flowing, natural) bodies of water, generally cold: "baptize . . . in living water. If you do not have living water, baptize in other water; if you cannot in cold, then in warm; if you do not have either, pour water on the head three times . . . ."[14] Only in some kind of emergency would baptism apart from such "living water" be practiced.

Yet in North American churches today this is the norm, with the effect that the practice of baptism is stripped of its original immersion into the larger interspecies context of a given place.[15] Such practice echoes symbolically the damming and pooling of once freely flowing waterways of all sizes across North America and the world, to catastrophic effect. As those who live near—and, much more so, those who used to live within—areas flooded in dam construction know, dammed water is deadly: not only destroying what it covers but also low in oxygen, high in pollution and sediment, and fatal to the complex interconnected cycles of flow, migration, spawning, soil and plant nutrition, and hydrology once sustaining the entire surrounding watershed. Laura Donaldson articulates these interwoven biological, hydrological, and human complexities, showing how contrary to life contemporary water commodification and damming is, and highlighting indigenous texts and traditions of alternative relationality with land and water.[16] Her coining of the term *aquacide* to describe such patterns of interlocking degradation and death is a shocking theological wake-up call. If the practice of Christian baptism requires immersion in *living* water, then Christians

14. *Didache*, chapter 7, cited in Bradshaw, *Early Christian Worship*, 8.

15. Exceptions include rural US African American churches and many communities in the global South that have retained the practice of baptizing in local bodies of water, and the Orthodox practice of Epiphany blessing of waters. On these see Hall, "Shall We Gather at the River," and Denysenko, *Blessing*.

16. Donaldson, "Covenanting Nature." On the urgent questions of access to clean local water especially in the Global South, see also Hofstad, "Murky Symbols"; Shiva, *Water Wars*; and Peppard, *Just Water*.

of most watersheds on Earth have work to do, difficult and slow, risky and economically challenging: to undam waterways, to cleanse them of waste and pollution, to reduce human over-consumption of fresh water, and to devise alternative economies of energy, agriculture, and industry currently threatening these baptismal flows.

Surely the steady separation of baptismal practice away from local rivers was a key symbolic step in Christian areas toward the alienation on which current patterns of exploitation and commodification of water, land, and humans rest.[17] Restoring and making "baptizable" every local creek, river, spring, or sea on Earth—that it too might be full of life, swarming with creatures depending on it—is thus a fundamental Christian imperative if new generations are to be baptized into such fullness of life, encountering in and near these waters countless forms each bearing the distinctive patterning of the living Word through whom it and all things were made (John 1:1–5).[18]

Local Waters

For living water is necessarily local water, *this* water in this waterway, filled with these particular species and nourishing those specific others. Caring for living water requires getting to know these particular others, by name, by sight and sound, by humility and care. It requires getting to know one's watershed as a whole: what impacts it, what threatens it, what helps and heals it. I have elsewhere argued for the importance of bio-regional grounding not just of Christian life (though certainly that) but of Christian ritual practice as well: that, in particular, baptismal renunciations and affirmations need to take forms specific to local bio-diversity and vocation, as well as orienting participants to the larger transcultural promises and claims of Christian faith.[19] A baptismal affirmation taking place in Alum Creek, with its Ohio diversity of freshwater shellfish, water plants, fish, insects, trees, mammals, etc., would sound and feel much different from a parallel rite taking place in my new home of Thousand Oaks, California, where the spring-fed (and runoff-polluted) creek running through Wildwood Park and over Paradise Falls nourishes a very different assortment of lizards, birds, insects, plants, and challenges. Baptizing in local waters also challenges the prac-

17. I support this claim in "Rewilding Christian Spirituality," 183–85. On the commodification of water see also Rasmussen, *Earth-Honoring Faith*, 266–78.

18. See McDougall, *Cosmos*.

19. Cf. Dahill, "Life in All Its Fullness." See too Myers, "From 'Creation Care' to 'Watershed Discipleship.'" Myers's work in "watershed discipleship" articulates this larger bio-regional orientation to Christian life itself.

tice of reliance on imported and/or bottled water, obliging Christians to learn to live within the limits of their watershed and to resist the practices of water commodification and profiteering that deprive millions of people worldwide of clean local water.

More recently I have been considesring how experiences like the one at the outset of this essay give rise to a fundamentally animist perspective. Graham Harvey uses the term *animism* "to refer to ways of living that assume that the world is a community of living persons, all deserving respect, and therefore to ways of inculcating good relations between persons of different species."[20] Getting to know my local watershed as a baptismal practice is inviting me into this animist vision precisely as an expansion of dominant forms of Christianity. That is, I understand baptism into local waters to mean immersion not only into the body of Christ (a human abstraction, albeit a meaningful one) but even more significantly into this larger sensory-physical relationality with the natural world: experiential kinship with creatures in the watershed of every phylum and kind, with local humans of diverse cultures and religions, and with natural forces and elements, all expressions of the *wild Logos*.

In this animist vision every creature and element is experienced as a real or potential Thou, the waters and watershed our sanctuary, our home, the wildness of it all the closest we shall come perhaps to the divine. Thus, out in the water, wanting to relate meaningfully and responsibly with the particular Thous I encounter there, I find myself immersed not (only) into the divine "persons" of the Christian Trinity but into the mutuality of personhood of *all these creatures* . . . and into what might be considered the personhood of the waterway itself.

Wild Waters

The language of the *wild Logos* in the preceding section moves to the heart of this essay, taking seriously the wildness of the waters into which baptism immerses Christians. I suggested just now that the language of divine

20. Harvey, "Introduction," 5. The language of "personhood" points obviously for Christians to the persons of the Trinity; I touch briefly on this connection shortly. But I would note here also the fact that Bonhoeffer too uses this language in a pivotal section of *Sanctorum Communio* (*SC*) where he articulates the I/Thou thinking of relations between individuals, the fundamental form of ethical relation: encounter with an other as a Thou is, he asserts, what makes an individual into a "person" (German: *Person*), obligated to the other by responsibility. See *SC* 34–57. I hear echoes of this Bonhoefferean vision of ethical personhood in the animist definition's interspecies view of such relationship.

"persons" as traditionally used in the doctrine of the Trinity needs amplification to include the much larger relationality into which baptism draws us, one that honors the "personhood" and subjectivity of the multitude of creatures who share our waters, creatures whose diverse and mysterious intelligences complement or surpass our own and whose wisdom and ways we urgently need again to respect and learn from in all humility. To consider these persons as among those into whom we are baptized, I suggest, makes the practice of Christian baptism both profoundly local and urgently relational, requiring us to learn patterns of receptivity and attunement from which privileged Westerners have for too long been cut off.[21]

But the encounter with wildness goes even further. Christians have long noted parallels between the great flood of Noah's time, destroying nearly all life on Earth, and the terrors of drowning implicit in baptismal metaphors and practice. Wild water is terrifying water, as survivors of raging floodwaters know, able to sweep away humans and our lives in an instant, and we do well to respect it. I've had my share of frightening encounters with wild water, from that childhood near-drowning to unexpected river rapids; and those vulnerable to rising seas and strengthening storms in this era of climate instability and chaos know much better than I do how destructive wild water can be to humans.[22] The wildness of a creek or ocean is part of its larger baptismal significance, including not only the force of the water itself but also the proliferation of its microbes in our guts from exposure to polluted water, the possibility of encounter with coyote or mountain lion or rattlesnake along these banks, the fact that creek edges are often home to humans on the edges of society.

I thus sense that this wildness—from which so much of our technology and so many of our walls buffer us—is in fact central to the encounter with the divine in baptism.[23] Rather than being baptized into Jesus Christ as a being

21. Many Christian texts describe contemplative attention to the natural world as a Christian spiritual practice. See, for instance, Christie, *Blue Sapphire*; Chase, *Nature as Spiritual Practice*; Dahill,"Bio-Theoacoustics."

22. For a tracing of religious dimensions of whitewater kayaking, see A. Sanford, "Pinned on Karma Rock." For insight on shifts in the planet's hydrological cycle due to climate disruption, see R. Sanford, *Storm Warning.*

23. Two essays I find particularly useful for pondering this nexus of human vulnerability, suffering, wildness, and mystery are Plumwood, "Being Prey," and Hatley, "The Uncanny Goodness of Being Edible to Bears." These connections make room for the possibility of divine encounter that is not safe—is even harsh or threatening—but require the caveat that I do not intend hereby to romanticize suffering let alone to divinize the effects of human injustice, sin, and irresponsibility in contributing to such suffering. Christians of the over-developed world especially need immersion into outrage against the human torment and species extinctions that fuel our privilege.

somehow conceptually separate from this place, I am coming to experience baptism as taking place into this larger wildness, that which contains and makes possible also my own. That is, I sense that it's precisely the wildness of the larger biotic world held in and sustained by the waterway in which I am immersed that in fact *is* its divine life, is *the* divine life of all that is. More than about a God separate from creation, therefore, I am curious about the wildness *of* each creature: a mystery that far transcends my capacity to know or explain it and a holiness in which I too may participate to the extent that I learn again how to speak and listen, give and receive, eat and drink, live and die fittingly within the larger wildness and mystery of the planet.[24]

Word and Words: Baptismal Formula

Surely one of the most vexing questions of theological language has to do with the verbal baptismal formula, specifically the Triune name with which initiates are baptized. The traditional formula, by which persons are baptized "in the name of the Father, the Son, and the Holy Spirit," emerges from key loci of the Christian Scriptures and early christological debates, and it provides a powerfully transcultural and cross-generational invitation into the tradition, including (unfortunately) the tradition's symbolic andro- and anthropocentrism. It is less successful, however, as language for inviting people into the mystery, physicality, wetness, wildness, and strangeness of the local watershed, the larger biosphere, or the encompassing cosmos into which a truly adequate baptismal practice must surely invite Christians today.

Thus I wish to think anew through the Book of Nature also on how to name the divine reality into which we baptize, hoping in the process to include enough distinctively Christian resonance to make such thought experiments at least conceivable to others. For this I find helpful Ellen Armour's invitation to consider *elemental* metaphors for the divine: Earth, air, fire, or water.[25] Her work uses Irigaray's category of the "sensible tran-

24. In "The Wild and the Self," Jack Turner notes that the OED definition of *wild* "gives the Latin equivalent *indomitus*, a word that suggests much about the subsequent use of wild" (23). This language raises the hair on the back of my neck, for it expresses what is most unorthodox about the thinking into which my experiences of living wild water are drawing me: namely that there's no *Dominus* out here, no superior incorporeal intelligence but just each new wild life, indeed the wild fullness of life within which each creature lives and dies according to its own best endowments, luck, wit, and improvisation.

25. Armour, "Toward an Elemental Theology."

scendental" to propose these elements as primary metaphors for the divine, useful for their physical/intuitive resonance available to all who live in the body and for their undermining of the anthropocentrism of much "God" language. Armour writes,

> Since the elements are not anthropomorphic in origin, they rather dramatically torque traditional Christian theism. Their particular combination of "is" and "is not," aptness and inaptness, in relationship to theism needs to be highlighted. Sensible transcendentals retain certain aspects of traditional Christian theism's insistence on divine transcendence: we are utterly dependent on them for our existence, and yet they exceed our grasp. Without air, for example, we literally could not exist, yet it does its work almost invisibly. The deepest parts of the oceans, the molten center of the earth, are literally inaccessible to us. The elements "are not" a God, properly speaking, yet they "are" that in which we "live, move, and have our being." Elemental theology, then, refocuses attention from an invisible, disembodied-but-agential transcendence to a (more or less) visible, embodied, impersonal transcendence.[26]

I'm intrigued with the use of "water" as a name of God at baptism, for the reasons Armour develops above. If we truly believe that in Christ God and the world are fully united and the world already permeated with all the holiness available to us, we need forms of ritual speech that reflect this one reality. The formula of a rite inspired by the holiness of Earth's actual waters might, for instance, name God-as-water here in this local river, in the Spirit/breath of all life, in the wild wisdom permeating the biosphere. "*Name*, I baptize you into Water: [*the Pacific Ocean, Lake Huron, Alum Creek*], holy Breath, the Life of the world."[27] If the traditional formula represents a baptism into the world opened by the Book of Scripture, then a version like this is an attempt to articulate baptism into the same world named through the Book of Nature. For Earth's waters polluted, depleted, hoarded, cherished, and wild *are* the "living waters" into which the Spirit immerses Christians.

26. Ibid., 53–54.

27. I make this proposal not to suggest that the scriptural and human-relational language and practices of traditional Christian baptism are somehow false or wrong: humans need touch and naming, human belonging, profound personal love, and the encompassing cross-cultural relationality traditional language provides. But humans also need intimacy and immersion in the more-than-human world, also filled with divine life—and this section explores what kind of language might evoke these connections ritually.

Being joined here to Christ incarnate, crucified, and risen inaugurates the conversion we need.

Conclusion

My encounters with living water, beginning with that kayak trip on Alum Creek, have borne me into what may seem an anomalous articulation of what Christian baptism means. Popular views of this sacrament tend to center in personal salvation, or larger human-communal belonging, or relationality with a deity fundamentally separate from (if intimately joined to, indeed incarnate within) the larger animal, vegetal, and mineral world. My immersions into living, local, and wild waters have broken open the unconsciously dualistic ways I too had previously thought of baptism. I am coming to view this sacrament instead as the invitation for Christians who worship this incarnate God—fully joined to the creation in *one* reality—to let psyche/spirit be reunited with our body's world, that watery life of all creation, and thus rejoined to the bodies of Earth itself, our heaven, our home.

Here in the water we meet all kinds of creatures, being at last restored to our own full humanity—for, as Abram reminds us, "we are human only in contact, and conviviality, with what is not human."[28] Here in the water we also meet lots of other humans, Christian or not, who share our dependence on these waters, and this encounter too invites us at last into the fullness of Christian faith—for surely, similarly, we are Christian only in contact, and conviviality, with those who are not. And finally, we meet here the sacred mystery, the ordering, intelligence, wit, and spark-fire of the vast creation itself, inviting us into the fullness of our gratitude (Latin *gratia,* grace) to past generations of Earth creatures who make up our flesh, into relationship with present creatures of all species and kinds, and into shared responsibility for future generations: that they too may flourish in these living, local, and wild waters.

BIBLIOGRAPHY

Abram, David. *Becoming Animal: An Earthly Cosmology.* New York: Vintage, 2010.

———. *The Spell of the Sensuous: Perception and Language in a More-Than-Human World.* New York: Vintage/Random House, 1996.

28. Abram, *The Spell of the Sensuous,* 23.

Armour, Ellen T. "Toward an Elemental Theology: A Constructive Proposal." In *Theology that Matters: Ecology, Economy, and God,* edited by Darby Kathleen Ray, 42–57. Minneapolis: Fortress, 2006.

Augustine. *De Genesi ad Litteram* [The Literal Meaning of Genesis]. In *Ancient Christian Writers,* edited by Johannes Quasten et al., translated by John Hammond Taylor, vols. 41–42. New York: Newman, 1982.

Barth, Karl. *Epistle to the Romans.* Translated by Edwin C. Hoskins. London: Oxford University Press, 1960.

Berry, Thomas. "Twelve Principles for Reflecting on the Universe and the Role of the Human in the Universe Process." *Cross Currents* 37 (Summer–Fall 1987) 216–17.

Bonhoeffer, Dietrich. *Ethics.* Edited by Clifford J. Green, translated by Reinhard Krauss et al. Dietrich Bonhoeffer Works (hereafter DBWE), volume 6. Minneapolis: Fortress, 2005.

———. *Letters and Papers from Prison.* Edited by John W. de Gruchy, translated by Isabel Best et al. DBWE 8. Minneapolis: Fortress, 2010.

———. *Sanctorum Communio: A Theological Study of the Sociology of the Church.* Edited by Clifford J. Green, translated by Reinhard Krauss and Nancy Lukens. DBWE 1. Minneapolis: Fortress, 1998.

Bradshaw, Paul. *Early Christian Worship: A Basic Introduction to Ideas and Practice,* 2nd ed. Collegeville, MN: Liturgical, 2010.

Chase, Steven. *Nature as Spiritual Practice.* Grand Rapids: Eerdmans, 2011.

Christie, Douglas E. *Blue Sapphire of the Mind: Notes for a Contemplative Ecology.* New York: Oxford University Press, 2012.

Clingerman, Forrest. "Reading the Book of Nature: A Hermeneutical Account of Nature for Philosophical Theology." *Worldviews* 13 (2009) 72–91.

Clingerman, Forrest, et al., eds. *Interpreting Nature: The Emerging Field of Environmental Hermeneutics.* New York: Fordham University Press, 2014.

Dahill, Lisa E. "Alive Together: Toward an Interfaith Interspecies Belonging." Forthcoming.

———. "Bio-Theoacoustics: Prayer Outdoors and the Reality of the Natural World." *Dialog: A Journal of Theology* 52 (Winter 2013) 292–302.

———. "Into Local Waters: Rewilding the Study of Christian Spirituality." *Spiritus* (Fall 2016) 141–65.

———. "Life in All Its Fullness: Christian Worship and the Natural World." *Liturgy* 31 (Fall 2016) 43–50.

———. *Reading from the Underside of Selfhood: Bonhoeffer and Spiritual Formation.* Princeton Theological Monograph Series. Eugene: Pickwick, 2009.

———. "Rewilding Christian Spirituality: Outdoor Sacraments and the Life of the World." In *Eco-Reformation: Grace and Hope for a Planet in Peril,* edited by Lisa E. Dahill and James B. Martin-Schramm. 177–96. Eugene, OR: Cascade, 2016.

———. "The View from *Way* Below: Inter-Species Encounter, Membranes, and the Reality of Christ." *Dialog: A Journal of Theology* 53 (Fall 2014) 250–58.

Dahill, Lisa E., and James B. Martin-Schramm, eds. *Eco-Reformation: Grace and Hope for a Planet in Peril.* Eugene, OR: Cascade, 2016.

Denysenko, Nicholas E. *The Blessing of the Waters and Epiphany: The Eastern Liturgical Tradition.* Burlington, VT: Ashgate, 2012.

Donaldson, Laura. "Covenanting Nature: Aquacide and the Transformation of Knowledge." *Ecotheology* 8.1 (2003) 100–111.

Francis. *Laudato Sí: On Care for Our Common Home.* Rome: Libreria Editrice Vaticana, 2015.

Gibler, Linda. *From the Beginning to Baptism: Scientific and Sacred Stories of Water, Oil, and Fire.* Collegeville: Liturgical, 2010.

Habel, Norman C. "Introducing Ecological Hermeneutics." In *Exploring Ecological Hermeneutics,* edited by Norman C. Habel and Peter Trulinger, 1–8. Atlanta: Society of Biblical Literature, 2008.

Hall, Taffey. "Shall We Gather at the River: A Photographic Essay of Believer's Baptism by Immersion." *Baptist History and Heritage* 45 (Winter 2010) 37–52.

Harvey, Graham. "Introduction." In *The Handbook of Contemporary Animism,* edited by Graham Harvey, 1–12. Durham: Acumen, 2013.

Hatley, James. "The Uncanny Goodness of Being Edible to Bears." In *Rethinking Nature: Essays in Environmental Philosophy,* edited by Bruce V. Foltz and Robert Frodeman, 13–31. Bloomington, IN: Indiana University Press, 2004.

Hofstad, Lynn. "Murky Symbols: How Contamination Affects the Symbolic Meaning of Water in Religious Rituals." Unpublished paper, Lutheran Women in Theology and Religious Studies/American Academy of Religion, 2015.

Johnson, Elizabeth. "Deep Christology: Ecological Soundings." In *From Logos to Christos: Essays on Christology in Honour of Joanne McWilliam,* edited by Ellen M. Leonard and Kate Merriman, 163–79. Waterloo: Wilfrid Laurier University Press, 2010.

Lathrop, Gordon W. *Holy Ground: A Liturgical Cosmology.* Minneapolis: Fortress, 2009.

Martin-Schramm, James B. "Bonhoeffer, the Church, and the Climate Question." In *Eco-Reformation: Grace and Hope for a Planet in Peril,* edited by Lisa E. Dahill and James B. Martin-Schramm, 110–24. Eugene, OR: Cascade, 2016.

McDougall, Dorothy. *The Cosmos as Primary Sacrament: The Horizon for an Ecological Sacramental Theology.* New York: Peter Lang, 2003.

McFague, Sallie. *The Body of God: An Ecological Theology.* Minneapolis: Fortress, 1993.

Myers, Ched. "From 'Creation Care' to 'Watershed Discipleship': Re-Placing Ecological Theology and Practice." *Conrad Grebel Review* 32 (Fall 2014) 250–75.

Nicholsen, Shierry Weber. *The Love of Nature and the End of the World: The Unspoken Dimensions of Environmental Concern.* Cambridge, MA: MIT Press, 2002.

Peppard, Christiana Z. *Just Water: Theology, Ethics, and the Global Water Crisis.* Maryknoll, NY: Orbis, 2013.

Plumwood, Val. "Being Prey." In *The New Earth Reader: The Best of Terra Nova,* edited by David Rothenberg and Marta Ulvaeus, 76–92. Cambridge, MA: MIT Press, 1999.

Rasmussen, Larry L. "Bonhoeffer: Ecological Theologian." In *Bonhoeffer and Interpretive Theory: Essays on Method and Approaches,* edited by Peter Frick, 251–67. International Bonhoeffer Interpretations Series. Berne: Peter Lang, 2013.

———. *Earth-Honoring Faith: Religious Ethics in a New Key.* New York: Oxford University Press, 2013.

Sanford, A. Whitney. "Pinned on Karma Rock: Whitewater Kayaking as Religious Experience." *Journal of the American Academy of Religion* 75 (December 2007) 875–95.

Sanford, Robert William. *Storm Warning: Water and Climate Security in a Changing World.* Calgary: Rocky Mountain, 2015.

Shiva, Vandana. *Water Wars: Privatization, Pollution, and Profit.* Berkeley, CA: North Atlantic, 2016.

Turner, Jack. "The Wild and the Self." In *The Rediscovery of the Wild*, edited by Peter H. Kahn Jr. and Patricia H. Hasbach, 27–50. Cambridge, MA: MIT Press, 2013.

The God(s) of November

—NATHAN KOWALSKY—

The encounter between spiritually attentive human beings and a wild animal dead at one's own hands is profoundly stimulating, both philosophically, theologically, and ethically. My father once told me that we went hunting because the world is broken by sin and so we have no choice except to live in it on its fallen terms. But that didn't explain why we were so happy to fit in with a predatory natural order that our theology said fell short of God's plan (if not being outright evil). Hunting with guilt and regret, or prefiguring the kingdom of God by eschewing hunting and meat eating altogether, both made more sense given the fallenness or sinfulness of nature. On the other hand, we could have been mistaken about God's plan for nature, and if so, then predator-prey relationships could be one of the sources for doing natural theology. I have followed this latter path, and so in this paper will briefly explore how my conception of God has been shaped by my being (among other things) a killer of deer. But first, I have to tell you what that's like.

I.

Every November I go hunting, and so I go home. I do not know how to hunt where I now live; there are too many unfamiliar landowners and trees, too much unlived ecology. But I know the grasslands, the prairies, the Great Plains of southeastern Alberta, the Bullshead Creek bed, the Great Hope Ranch, Uncle Herman's old farmstead, that smell of that sun-warmed grass, that species of brush where mulies bed down, that wind gnawing at my face.

I don't live there anymore, but it will always be home. My soul was formed by (among other things) hunting in the folds of that land, and hunting is now the only real chance I have to return, even if for a moment, and be something other than a visitor. From the outside, going hunting in November looks like a leisure activity (which, technically, it is), but from within it feels like anything but. It is the living of a different lifeway, a procurement activity with products that last the year round, and a passage through a window into a world that remains, the rest of the time, behind closed doors and obscured from the vast majority of the cloistered West.[1] I live in the cloistered West, but that's not where my home is.

In this story, so far, the animals are not really there. If you go hunting to see wildlife, you'll likely be disappointed. All the more so if you go hunting to see killing: the majority of hunts are unsuccessful.[2] As José Ortega y Gasset points out, "hunting is not simply casting blows right and left in order to kill animals or to catch them. . . . Rather, the fundamental task of all hunting [is] bringing about the presence of the prey."[3] Presence can be an odd notion; it is a kind of existing that requires a subject to be present *to*. (If there isn't anything to be present to, then there is no presence, no presentation, of the thing.) Even the word *existence* itself comes from the Latin *existere*, "to stand forth." When something stands forth out of nothingness, it exists, is apparent, is an apparition. The hunted animal does not simply stand out. Hunting is *seeking* an apparition, seeking the existence of something that does not yet, or may never, exist. Hunting seeks a "being there" when there is apparently no being there.[4] It is all too often forgotten that hunting *means* searching, and waiting, and not always killing. Before the killing comes the animal, and before the animal comes the search for something that is not present.

1. I use the phrase "the cloistered West" to refer to the separation and isolation of so many urban and rural cultures from the ecosystemic processes of wild landscapes (not to monasteries necessarily). I also recognise that this segregation is not limited to Western cultures.

2. Government of Alberta, "Recreational Access Management Program Pilot Study Year 1 Report."

3. Ortega y Gasset, *Meditations on Hunting*, 75, 76.

4. I am using phenomenological language here (as might be expected, given how much Ortega has in common with Heidegger), but my point is simple: from the perspective of the hunter, there are not any animals out there until they are present ("there") *to* the hunter. Brian Seitz says that going "home empty-handed [is] a universal experience familiar to any hunter." "Hunting for Meaning," 73–74. If no animals present themselves to the hunter, then the experience of hunting is of emptiness or nothingness vis-à-vis the existence of the animals sought. (I have written at more length about this theme in Kowalsky, "The Waiters.")

If there are animal others out there, in the nothingness beyond presence, then they are, for the most part, hiding. They are not simply being hidden by something else; they are using something else to hide themselves. They are using the land as cover. They are active subjects actively hiding from beings like me who seek to find, capture, kill, eat, and use them. Moreover, I am not the only being they are hiding from. There are other beings like me, nonhumans, that also seek out this prey, and so, in this respect, I am like these other animals, the predators. They too are hidden, for similar though inverse reasons. The prey are watching for predators—with eyes, ears, and noses—preparing to evade them should they be found. Thus the predators must be hidden too, and use the land to cover themselves if they are to get close to the prey. This is why, for me, hunting is a way of walking, standing, crawling, smelling, waiting, and looking. If you do not do these things well, the prey will not appear.

Likewise, being hunted is a way of walking, standing (or sitting), crawling, waiting, listening, smelling, and looking. If you do not do these things well, the predator will appear. While hunting is "a contest or confrontation between two systems of instincts,"[5] both predator and prey are alike in their hiddenness and their searching. I, a human hunter, thus experience this world as a participant among participants. They seek signs; so do I. They make themselves scarce; so do I. We all use the land to cover ourselves (with distance, defilade, brush, silence, stillness, slowness, camouflage). Each of us, in our own ways, seeks to uncover the hidden being of our concern. Perhaps this is why some call hunting a game, or call prey "game animals." It is a form of play, or interplay, but this does not make it a sport. Such play is simply the way of life of wild animals. Thus does Ortega claim that hunting places humans "within the orbit of animal existence" and that hunting leads us "to adopt the attitude of existence by which wild animals generally live."[6] We are not radically different, the wildlife and I, predator or prey.

Of course these animals are subjects. Hunting wouldn't be possible if they weren't—to hunt well is to be able to predict what your prey is thinking, or to put oneself, as it were, into their shoes. This cannot be done with Cartesian automata.[7] So, on a cold November prairie morning, what are the mule deer thinking (assuming there are any)? They could be anywhere, but if it is windy they will probably be in the lee of a hill with their backs to the wind, smelling what is behind them and on the lookout for what is in front

5. Ortega y Gasset, *Meditations on Hunting*, 64.

6. Ibid., 121, 137.

7. Cf. Kover, "Flesh, Death, and Tofu," 174: "the hunter must constantly imagine the *subjective* state of her quarry, attempting to quite literally perceive the situation through its senses."

of them. (This is why I would be a bad hunter anywhere else, having to learn how an entirely new ecology interacts with different species' behaviors.) If the land has been overrun with truck hunters, then (if I'm lucky) the deer will be hiding off and away from the tire tracks left by tractors and half-tons. Let me say it again: *they could be anywhere* and *if I'm lucky*. Wherever the prey animal might be, it is not there for me. Rather, it is where it is for itself. It is free. This is why my hunting search carries me off roads and over fences, because the hidden beings I seek are not tethered and, to some extent, neither am I. The search for the hidden is the life of the free animal, nonhuman or otherwise. What is the prey animal doing? It is certainly doing what it does for itself, not for me, although it may be doing that on account of me. Tethers, fences, cages, bridles, and house training make demands of animals, ensuring that they are there for (certain) humans. Such animals cannot be hunted unless they are untrained, untethered, and freed from such restraints. But I cannot make a free animal be there for me; I cannot even know if there might be one there for me. All I can do is be there and be ready in case there is one for me, but I do not reveal it. I do not uncover it from beneath the folds of the land. I must put myself in its path (not knowing if it is there to have a path), hoping that it will break cover when I pass by or glance its way, but none of this is under my control. In this sense, the animal I hunt is wild. Wildness means "willed," to be under the control of one's own volition. But animals under human control are domesticated, under the lordship of the domicile and the *dominus*, dominated. Not so the hunted animal. It is under its own control. All hunted animals are free, and all free animals are hunted, predator or prey alike.

If the animal presents itself to me, only then can I reach out and attempt to capture it with a weapon. I am certain that the animal does not want me to do this. That is, as noted above, why it was hiding in the first place. The prey animal does not wish to be present to me, to be found, or for me to be on its path. It does not wish to be caught or touched. It does not wish to die, to leave its body to become my food or clothing. Nothing which hides desires this for itself, myself included. In this we are of like mind. Yet out there, each November, we both hide and seek to uncover the other's presence. Up until this point, the sought animal has been generic, not particular, but now, as we near the prey in the flesh and not just in thought, the individual appears. Here it gets personal. That animal, *there*, does not wish to be caught—that mule deer doe, rising above a bed in the scrub to see who disturbed her, that one which my bullet missed, who ran into the cover of the land again—but another unwilling mulie, later, where my skill was sufficient this time, this one felled by my bullet, this one that I have killed. This one, and not any others. This one is the individual I have found, and such individuation is lethal. In catching her (or him, one does not always hunt does), I have divided her life from her body, and will divide her skin from her meat and her meat from her viscera and bones. Individuals are not indivisible. But in catching her, her individuality has been lost just as it has been found, for in death the living breath vanishes to I know not where. Even here, at the very moment hunters seek, the prey is not under control, for it has escaped. But control is not what hunters seek. They seek the enchanting animal, find it only in the paradox of death, and take only the trace, which is the body.

I must speak rightly of my deadly encounter with the individual wild animal; to do otherwise would dishonor these animals, and this one in particular. It would also dishonor the tradition of hunting received from my forebears, the patrimony of the human species stretching back hundreds of thousands of years before the irruption of agrarian and civilized Earth mastery. Each animal fallen by my hands is different—though almost always a deer, and almost always a mulie. Each one has its own story, and I know little of it. Hunts, as I have said, are silent and slow affairs, but the attempt to kill is loud, occurring in a split second. Adrenaline skyrockets; I run and shout, and my ears ring from the sound of gunshot. But almost as quickly, the stillness returns, and (if I have caught one) I am with a wild animal which, were it alive and that close, would kick and bite and gore me nearly to death. This doe is warm beneath my hand. Her eyes are not yet glassed over. Her fur is smooth. There is little blood. When I reach her, she continues to breathe, and though I know she is in massive shock, possibly inuring her to the pain, and will bleed out in a few more seconds, it feels wrong to let her wait it out. I cut her throat; I hasten the end. My grandfather taught me how. I do not like to do it, but feel that I should. If I were in her place, I would want to be put out of my misery. I give her a mouthful of grass as a symbolic last meal, a ritual I learned from the Germans. She does not look like me and does not have to. Despite our differences, we share much: round brown eyes, two ears, a nose, sharp teeth, a pink tongue, hair, four limbs. I cut her open, eviscerating her body to let her flesh cool down. I begin at the bottom of the gastrointestinal tract. I remember being embarrassed at first, but as my father pointed out, we all have an anus and sexual organs; this is how God made us. My father was right, and some day I too will be dead and naked on a slab being "prepared" by strangers. I cut her skin straight up to her throat, and even then there is still almost no blood. I saw through her pelvis and rib cage, and then cut her esophagus and trachea away from her backbone, continuing down to her heart and lungs, and diaphragm and liver, and stomach and intestines, all of which come out in one big interconnected piece. I see what George Bataille means when he says we are all basically "tubes with two orifices"—everything else is just an appendage.[8] Inside the remaining cavity is where all the blood is, congealing already. I tip her over and leave a feast for all the prairies' scavengers, who by morning will have left only a faint red smear behind. Hopefully by now someone has gone and brought a truck over, so we only have to pull her for a few hundred meters (and hopefully not up too steep a coulee). Then it is time to rest.

I have never seen my own trachea, and I hope I never do. But I have seen hers, and I am sure it is close enough to mine. I have seen the "green fire" in the eyes of Aldo Leopold's wolf also die in my doe's eyes, and I imagine mine might go out like that someday too. In short, I see my own mortality in hers. There is no triumph in her death. I see, rather, my future in her present, and it is a serious business, facing one's own mortality in the flesh of the world. Perhaps this is why we pray over our deer's bodies, my family and I. We are encountering the limits of our own lives in the ending of these lives, and we need to recognize the presence of the god(s) in such doings. Sometimes we seek and find, sometimes we do not find. Sometimes we are found. For a time, we live. At the end, we die. In any death, another feeds, another lives. This time I was the one fed, but someday other critters will feed on me. There is no conquering this death, no escape from this trap, not even in the cloistered West. All I can hope for is that life will always win over and above it all. That is the substance of many faiths.

Can we now say, as the psalmist does, that "This is the day the Lord has made, let us rejoice and be glad in it" (Ps 118:24)? Facing one's own mortality every November is not tragic or depressing, after all. It is just sobering, and beautiful always. I am not any less at home because of it. Indeed, it would be strange to say that because of my encounter with my own mortality in the death of my prey, I am not at home in the wild land where my prey is found and where I find myself each November. We are in it together, which is to say that we belong here on account of it. It is a way of life to die and to feed this way, and it is a life worth being loved. So we keep on hunting. We take our deer to the house and hang them up for skinning. A few days later, we take them to our butcher, who makes amazing double-smoked Mennonite farmer's sausage. I do not need to buy any more red meat for a whole year. We will pray over these meals as per usual, but in blessing the hands that have prepared the food, I will also be thinking of hooves. I am learning how to make rawhide and parchment from deer skins, and I can take them to the tanners to get leather for apparel. This deer, or that one, is in my body or on my body or in my house. I can show you the photos, or tell you the story of

8. Bataille, *Visions of Excess*, 88.

this buck whose antlers memorialize him to me.[9] Dead deer permeate our little life here in the city, but my real home is where the deer live. I do not belong in the city, and I suspect that, really, none of us do. For people of my ethnic stock, home actually was—ten thousand years ago—where the deer lived. For others, it was that way only one hundred years ago, or maybe still is that way for some. That way of living and being at home embeds a person in a particular landscape ecology as a participant alongside other subjectivities with whom one is metaphysically consanguinous in death and food and clothing. Each type does their own thing, human, or cervid, or lupine, or otherwise, but all are seeking and finding, living and dying, and life keeps on going for evermore. It'll do.

II.

Every November I go hunting, and the god(s) are there before me. I am a Trinitarian Christian, which must be why I'm comfortable with a plural monotheism: hear, O Israel, the one is many.[10] In Annie Dillard's *Holy the Firm*, each day has its own image of God.[11] I see this as sacramental: the god(s) are only known to us creatures by means of creatures themselves. All creation sings God's praise, and all spirit is mediated through the flesh. So what might the god(s) of the hunting day be like? What spirit is mediated through that flesh? When the sun rises over a rolling land encrusted with fresh snow, painted pink and white like a marble baroque cupcake bakery but immense in expanse, vast in silence and cold, the sky soaring overhead like the bent bow of Earth itself, I am not alone in swearing that this is God's country. My father worships here more than in any cathedral. But for all that the god(s) do not stand forth. They are hidden by the land just as the deer are (and as the hunters try to be), so what are we to make of them?

In this question the well-traveled trope of the hiddenness of God arises. But it is important remember that this hiddenness is not an agrarian trope, not something analogous to animal companionship or domestication. The hidden god is a wild god because it hides. It is not without reason that Robert Farrar Capon calls theology "hunting the divine fox."[12] There

9. Cf. Marvin, "Living with Dead Animals?," 110–12, 115–16.

10. I will use the capitalized singular "God" for the personal name of the deity of classical monotheism, and the lower case occasionally plural "god(s)" for less specific or more inclusive gestures towards divinity in exploratory natural theology.

11. Dillard, *Holy the Firm*.

12. Capon, *Hunting the Divine Fox*.

is no way to know if there is a divine fox out there to hunt; a wild god may very well not be there at all. Atheism is always possible. If you want sight rather than faith and doubt, a tamed pet god is for you. If you keep it on a leash, you will always know where it is and that it is there for you. This is the god of philosophy, the deity which appears "only insofar as philosophy, of its own accord and by its own nature, requires and determines that and how the deity enters into it." But as Heidegger goes on to say, "Man [sic.] can neither pray nor sacrifice to this god. Before the *causa sui*, man can neither fall to his knees in awe nor *can* he play music and dance before this god."[13] On the other hand, an unleashed god can be worshipped because it exceeds our grasp and control. It is a Spirit that blows where it wills, like the wind which gnaws my face in November. I do not have time for the useless gods of caged days.

This makes God difficult to catch, as if God is not there for me. Is theology supposed to capture God? If so, what would we rightly do with that quarry? God could be handled and released (like a fish), leashed and caged (like a horse), or killed (like a deer). None of these options seem right, although in the Christian religion, God *is* killed but with uncharacteristic effects. The irony, then, may be that if you really want a dead God in the Nietzschean sense, a leash or a hook is the way to go. But a theology of the hunt never achieves mastery of the divine, just as a hunter never masters the prey, for even in death it escapes and leaves only the trace of its body.[14] What's more, the body is what gives life to the scavengers and the hunter's community, which is all the hunter sought in the first place. Perhaps this is another reason why I see monotheism in the plural: no appropriation of the divine will ever capture its aseity, its being-in-itself.[15] All we can hope for are traces that nourish us and a faith that helps us with that. While there is something eucharistic in this model, let us not get carried away: I know that the deer isn't giving itself to me freely as Christ went to the cross,[16] and what's more, hunting is not sacrificial because it is not a form of economic

13. Heidegger, *Identity and Difference*, 56, 72.

14. Here I refer the reader back to my earlier point that the prey animal's "individuality has been lost just as it has been found," because death removes any possibility for controlling or dominating a living being.

15. The Oxford English Dictionary (1989) defines aseity as "underived or independent existence," first attested in English in 1691 with a Latin etymology: *ā* ("from") and *se* ("oneself").

16. In saying so, I am stepping outside the phenomenological perspective and assuming a sympathetic anthropomorphism to identify the subjective state of the animal. Phenomenologically, the appearance of the animal *is* a gift to the hunter. Cf. Ingold, *Hunters, Pastoralists and Ranchers*, 282–84. I cannot, however, explore this intersection further here.

bargaining with the gods.[17] So if the god(s) are hidden by the land, I do not think there's any way of flushing them out. They are not like deer that way. And if there are no god(s), not even my death would reveal this truth to me. But they are like deer in their freedom from domestication and domination, and only in that freedom are the god(s) meaningfully divine. If the land lacked such divinity, it would not feel full enough. And yet when I hunt, the land and I could not possibly feel any more full. So, I have to think the divine is in there somewhere, deep down things.

All this is a world, I submit, to be at home in. Wilderness was our species' home for at least 90 percent of its timeline, and so it is odd that so many of us are not at home there. Maybe it is because of the killing. How could we forget that when the god(s)—who are not there for us—nourish us with the flesh of the wild world, their gift of nourishment comes via death? That's not particularly decent; it's why civilized folk often call hunters barbaric or savage (even though those terms are xenophobic). It is also why hunting is seen as old fashioned, unprogressive, and—ironically, at a time when we all "love animals"—as debasing people to the level of "beasts" who cannot choose to behave otherwise. It is not good to call someone a "predator." It is not seen as good to be at home outside the cosmopolitan city, or to deal in blood. Hunting is uncomfortably comfortable with the way wild land is, as if the land is also the way it should be. Hunters appear unconcerned that (according to David Hume and G. E. Moore) to equate "is" with "ought" is a fallacy, and that nature is (according to J. S. Mill) the worst possible example of good human behavior.[18] Hunting is comfortable with endemic mortality and a lack of cultivation; one might as well be okay with murder and ignorance. Holmes Rolston, III suggests that hunting is a sacramental affirmation of the way the world is made,[19] but the god(s) of that world must be bloodthirsty and cruel.

Yet this is the cloistered West talking, where treasured lives (including "livestock") are protected by barriers from the murderous chaos that supposedly lurks outside. Inside are the gardened plants and teacup poodles and cooped-up chickens, while outside are the weeds, the fleas, and the vermin (i.e., all of the untreasured lives) that would steal away the lives we protect, as if true life were a jewel, a beautiful but static and hard stone that can retain its luster for all eternity if it is tended properly and locked away from thieves. In the carcass of the doe—and the wild land she signifies—I

17. Cf. Kover, "Of Killer Apes and Tender Carnivores."

18. Hume, *Treatise of Human Nature*, §3.1.1.27; Moore, *Principia Ethica*, Kap. 1; Mill, "Nature."

19. Rolston, *Environmental Ethics*, 91.

encounter a god that is other to the putative comforts and safeties of the walled city, its fenced fields, and its conception of everlasting life. Capon says his God is "inordinately fond of rough places"[20] and this fondness should be clear enough from natural theology. If God brings the world to be out of her/his love for it, then the love of God is first for wild things that scoff at the commotion of the town and the driver's shout (cf. Job 39:7ff). Moreover, revealed theology tells us that the world God brings to be through love is good, very good (Gen 1:31), and we know from natural history that no human sin has broken biology in any fundamental (ontological, theological) sense to become less than very good. And so it seems to me, a hunter, that the goodness of the world is the wildness of God, and that good world is the wild land the hunters says "yes" to. There is nothing about the wild world which the hunter needs to change; any changes turn hunting into farming and ranching (which is not to deny that many hunters are farmers or ranchers in their day jobs). The hunters, along with the deer and all the animals around, are part of a flow, a way of life that they participate in rather than resist or reject. It is rather the roads, the fences, and the towers of steel which say to that land "you will not do, you will be 'developed to suit,' you will be tamed, mastered, and controlled." That these endeavors never fully succeed is beside the point; it is rather the thought that counts.

So, like St. Augustine in his *Confessions*, the hunter in her or his actions says "Far be it then that I should say, 'These things should not be.'"[21] My dad was right that we hunt because to do so conforms to the terms of the world's natural order, but he was wrong to think that those are the terms of sin. They are rather the terms of the god(s) of November, and these gods are other to the terms imposed by the domestic, sedentary cultures of some humans (including the culture I live in). This sort of divinity is numinous, unfamiliar to that culture, and transcendent rather than ontotheological. There may be terror before it, but the hunter (at least) is nourished by the grace it gives to quench fear.[22] So while this recalls the well-worn trope of the Otherness of God, we would do well to realize what this god is other to. It is other to the hunter, of course, and other to the animal—the animal is neither sacred in itself nor a sacrifice—but it is a sacrament of the wild land that hides the god(s) that are other to the cloistered West, where life is thought to be kept safe like an immortal jewel. Out here, in the land which environs even those cultures, life is not a thing to be had but a process to be lived. Living is not the absence of death and death is not the end of life,

20. Capon, *The Third Peacock*, 16.

21. Augustine, *Confessions*, §7.13.

22. Cf. Otto, *The Idea of the Holy*, 24.

rather life is a continual interdependence of dying and coming alive. This is the eternal life, I suggest, of the carcass of the doe. I quote at length from the Benedictine monk David Steindl-Rast:

> Life, if it isn't a give and take, is not life at all. . . . It must be stressed that this is not an either/or; life is not a give or take, but a give and take; if we only take or only give, we are not alive. . . . Whenever we give ourselves to whatever presents itself instead of grasping and holding it, we flow with it. We do not arrest the flow of reality, we do not try to possess it, we do not try to hold back, but we let go. . . . And whenever we do give up a person or a thing or a position, when we truly give it up, we die—yes, but we die into greater aliveness. We die into real oneness with life. Not to die, not to give up, means to exclude ourselves from that free flow of life. . . . So in many traditions you have this notion that throughout our lives we train for a right dying; and that means to train for flowing with life, for giving ourselves. . . . Then comes the moment of death, whether it is the ultimate death or a moment in the middle of life, and we give up our independence and come to life in interdependence, which is the joy of be-longing and of being together.[23]

For me, hunting in November is a training for death, and I am never more alive than then. To be another dying animal living among dying animals means I belong to the wild land more so than if I worked that land or owned it. Hunting is a giving up on the attempted permanence of the domicile, and an opening up to a far deeper belonging to the god(s) and their land which, to the surprise of many, is not hostile but life-giving in and through (among other things) the bodies of deer. In this coming to life in interdependence, the god(s) are left free. They aren't even captured at all. There is no having them, even in death. Rather, in death they are glimpsed as alive, and that is all the having of them I think we should expect.

Bibliography

Augustine. *Confessions*. Translated by R. S. Pine-Coffin. London: Penguin, 1961.

Bataille, Georges. *Visions of Excess: Selected Writings, 1927–1939*. Translated by Alan Stoekl. Minneapolis: University of Minnesota Press, 1985.

Capon, Robert Farrar. *Hunting the Divine Fox: An Introduction to the Language of Theology*. San Francisco: Harper & Row, 1985.

23. Steindl-Rast, "Learning to Die," 22–31.

———. *The Third Peacock: The Goodness of God and the Badness of the World*, rev. ed. San Francisco: Harper & Row, 1986.

Dillard, Annie. *Holy the Firm*. New York: Harper & Row, 1977.

Heidegger, Martin. *Identity and Difference*. Translated by Joan Stambaugh. New York: Harper & Row, 1969.

Hume, David. *A Treatise of Human Nature*. London: John Noon, 1739.

Government of Alberta. "Recreational Access Management Program Pilot Study Year 1 Report."

Ingold, Tim. *Hunters, Pastoralists and Ranchers: Reindeer Economies and their Transformation*. Cambridge: Cambridge University Press, 1980.

Kover, T. R. "Flesh, Death, and Tofu." In *Hunting—Philosophy for Everyone: In Search of the Wild Life*, edited by Nathan Kowalsky, 171–84. Malden, MA: Wiley-Blackwell, 2010.

———. "Of Killer Apes and Tender Carnivores: A Shepardian Critique of Burkert and Girard on Hunting and the Evolution of Religion." *Studies in Religion/Sciences Religieuses* 46.4 (December 2017) 536–67.

Kowalsky, Nathan. "The Waiters." *Orion: The Hunter's Institute*, September 16, 2013. http://fairchasehunting.blogspot.ca/2013/09/the-waiters.html.

Marvin, Garry. "Living with Dead Animals? Trophies as Souvenirs of the Hunt." In *Hunting—Philosophy for Everyone*, edited by Nathan Kowalsky, 107–18. Malden, MA: Wiley-Blackwell, 2010.

Mill, John Stuart. "Nature." In *Three Essays on Religion*, 3–65. New York: Henry Holt, 1874.

Moore, G. E. *Principia Ethica*. Cambridge: Cambridge University Press, 1903.

Ortega y Gasset, José. *Meditations on Hunting*. Translated by Howard B. Westcott. Belgrade, MT: Wilderness Adventure, 1995.

Otto, Rudolf. *The Idea of the Holy*, 2nd ed. Translated by John W. Harvey. London: Oxford University Press, 1950.

Rolston, Holmes. *Environmental Ethics: Duties to and Values in the Natural World*. Philadelphia: Temple University Press, 1988.

Seitz, Brian. "Hunting for Meaning: A Glimpse of the Game." In *Hunting—Philosophy for Everyone: In Search of the Wild Life*, edited by Nathan Kowalsky, 69–79. Malden, MA: Wiley-Blackwell, 2010.

Steindl-Rast, David. "Learning to Die." *Parabola* 2.1 (Winter 1977) 22–31. http://parabola.org/2016/02/29/learning-die-brother-david-steindl-rast/.

Ecological Transformation through Attentiveness and Intimacy

—CRISTINA VANIN—

Experiences of an Ecological Spirituality

While I was working on a master of divinity at the University of St. Michael's College in Toronto, I became an associate of Holy Cross Centre for Ecology and Spirituality, a retreat center run by the Passionist community of Canada and situated on the shores of Lake Erie, Ontario.[1] In the 1970s, the community decided to shift the focus of the retreat work toward ecology and spirituality because of their encounter with cultural historian and geologian, Thomas Berry. It meant a shift in the kinds of programs that were offered, and in the way the center itself functioned, that is, with greater attentiveness to the local bioregion, its use of water and energy, growing its own food, composting, etc. It also meant a shift in spirituality as the community tried to integrate Passionist spirituality that is focused on the way of the cross, with the new story of the universe.

One of the ways in which this deeper, broader, and more comprehensive spirituality was embodied at the Centre was through the building of a pilgrimage walk on the property called "Stations of the Cosmic Earth." It celebrated eight moments of grace in the story of the universe: the emergence of the universe, of the Earth, of life on Earth, of the human, of

1. The Passionist order was founded in 1720 by Paul of the Cross, "who saw in the Passion of Jesus 'the greatest work of divine Love' and the revelation of the power of the Resurrection to overcome the forces of evil." Passionist spirituality is focused on the way of the cross and the passion of Jesus, the suffering that Jesus underwent on the way to his death.

agriculture, of culture and religion, of science and technology, and of the newly emerging Ecozoic Age.[2] It was meant as a pilgrimage that took you into the story of the universe, the joys and beauty of its ongoing emergence. But it was also meant as a pilgrimage into the increasing suffering of the Earth. I often heard Thomas Berry and the staff at Holy Cross talk about the passion of Christ as the passion of the Earth. Through many celebrations of the Easter Triduum at Holy Cross, I came to appreciate that the way of the cross can now mean walking with Jesus as he suffers with and in the Earth today. These same eight stations were captured in stained glass windows in the chapel at the Centre. In addition, if you visit St. Gabriel's Passionist Church located in Toronto, you will find a series of stations of our cosmic earth located throughout the garden that is in front of the church, stations inspired by the original chapel windows.[3] In 2004, I had the opportunity to visit the Maryknoll Ecological Sanctuary in the Philippines, which has developed its own meditative walk on the universe story, one that integrates local indigenous crafts and images. In 2009, the Ignatius Jesuit Centre of Guelph, Ontario, developed a set of twenty-five stations of the cosmos.

Every opportunity that I have had to walk one of these sets of stations reminds me that we need to give shape and meaning to our human lives by relating who we are to the story of the universe. Berry writes, "Our sense of who we are and what our role is must begin where the universe begins. Not only does our physical shaping and our spiritual perception begin with the origin of the universe, so too does the formation of every being in the universe."[4] I continually find myself facing a question that Berry consistently asked: What is it about the way in which we understand ourselves that makes us wreak such havoc on the planet without seeming to be very affected by the consequences of our actions? I am challenged to see that there are problems with the way I understand myself and the natural world. Over many years I have slowly awakened to understanding myself as part of an amazing universe story.

Such a shift in understanding the nature and role of human beings requires us to do more than appreciate nature if, in the meantime, we still think and understand ourselves as quite distinct from it, even quite separate from it. Instead we need to enter deeply into the dynamics of creation and begin to know that the story of the universe is also our own story. It is about learning that each moment of the story is a graced moment calling me to deeper intimacy and transformation.

2. See Swimme and Berry, *The Universe Story* and the Big History Project.

3. http://stgabrielsparish.ca/who-we-are/green-church/engineering/

4. Berry, *The Great Work*, 162.

The Need for an Ecological Spirituality

In his essay "An Ecologically Sensitive Spirituality," Thomas Berry argues that what is most urgently needed is a reorientation of all aspects of human living "toward an intimate experience of the world around us."[5] Berry suggests that such a reorientation will need the help of "a spirituality that emerges out of a reality deeper than ourselves, a spirituality that is as deep as the Earth process itself, a spirituality that is born out of the solar system and even out of the heavens beyond the solar system."[6] Such a spirituality helps us to recover a capacity for being in communion with the Earth and understanding ourselves as integral with the universe process.

This is where the world's religious traditions have their role: they can help us appreciate that the story of the universe has a dimension to it that transcends the physical, and that the universe, from its beginning, is a psychic and spiritual as well as a physical reality. Christianity has its own capacity to contribute to the transformation into a larger, more comprehensive, and deeply spiritual realm of being because "[o]nly religious forces can move human consciousness to the depth needed. Only religious forces can sustain the effort that will be required over the long period of time during which adjustment must be made. Only religion can measure the magnitude of what we are about."[7] If Christianity is to help human beings to establish a relationship of reciprocity and intimacy with the rest of natural world, it will have to be transformed into a new cosmological context.

Berry warns that such a change "is not possible, however, so long as we fail to appreciate the planet that provides us with a world abundant in the volume and variety of food for our nourishment, a world exquisite in supplying beauty of form, sweetness of taste, delicate fragrances for our enjoyment, and exciting challenges, for us to overcome with skill and action."[8] And, at its root, this is the most serious difficulty that we are facing. Much good work is being done in education, economics, agriculture,

5. Berry, *The Sacred Universe*, 132.

6. Ibid, 74.

7. Berry, *The Christian Future and the Fate of Earth*, 11. See Bernard Lonergan on religious conversion in which "[a] total being-in-love [becomes] the efficacious ground of all self-transcendence, whether in the pursuit of truth, or in the realization of human values, or in the orientation [human beings adopt] to the universe, its ground, and its goal." Furthermore, any religion that helps human beings to develop "to the point, not merely of justice, but of self-sacrificing love" can help to sustain us through the sacrifices that will be required of us as individuals and as a species as we respond to the ecological crisis. Lonergan, *Method in Theology*, 240–41. See also, Ormerod and Vanin, "Ecological Conversion."

8. Berry, *The Sacred Universe*, 48.

global governance (the United Nations Earth Charter), religions—including Christianity, and conservation.[9] But Berry insists that all of these efforts do not seem to be enough because we continue to devastate the Earth in so many ways: climate change, deforestation, loss of biodiversity in species and cultures, pollution, waste, the list goes on.

> The magnitude of the devastation demands more than the series of reforms we have cited. It calls for a change of mind, for universal acknowledgement that the human is a subdivision of Earth and as such bears a responsibility for Earth's health. It calls for global commitment to assist the planet in the recovery of its vigor.[10]

Berry identifies three things that make it difficult for Christianity to assume adequate responsibility for the fate of the Earth community: the loss of the Book of Nature, an emphasis on redemption *from* the earth, and a focus on the soul.

The advent of the printing press in the sixteenth century made written Bibles more available so that the earlier Christian tradition that acknowledged two interrelated sources of revelation—the natural world and the biblical world—diminished. As the Book of Nature disappeared, Berry suggests that Christians became increasingly more "hesitant to enter profoundly into the inner reality of the created world in terms of affective intimacy."[11]

A second difficulty is that Christianity has an emphasis on redemption *away* from this world, which is considered flawed, seductive, spiritually irrelevant. "We are here, as it were, on trial, to live amid the things of this world but in thorough detachment from them. We long for our true home in some heavenly region."[12] A significant consequence of this redemptive spirituality is the development of a deep barrier to our being able to become an intimate presence to the rest of creation. Furthermore, the affirmation of the divine as the transcendent, personal, creator of all has also made it difficult for Christians to feel pulled toward establishing an intimacy with Earth as they do with God.[13]

9. See Berry, *The Sacred Universe*, 37.

10. Ibid., 166. Berry refers to prominent biologists who are promoting this view of the Earth, such as E. O. Wilson, Peter Raven, Paul Ehrlich, and Norman Myers.

11. Berry, *The Christian Future*, 38.

12. Ibid., 39.

13. See ibid., 6: "The salvific, redemptive traditions of the West tend to save humans out of the temporal order or to assign meaning to the temporal order in terms of a 'salvation history,' with an ultimate goal outside of time. The emphasis is on trans-temporal experience."

Finally, Christianity emphasizes a human soul that was created directly by God.[14] This situates the human as above or apart from the rest of creation, and given the responsibility to care for all that has been created. However, we have become ambivalent towards the natural world. We have lost an understanding that there is a primary and inherent relationship between the human and the divine in the natural world itself, which has affected how we exercise our responsibility.

> But this leaves us in an alienated situation. We become an intrusion or an addendum to the natural world. Only in this detached situation could we have felt so free to intrude upon the forces of the natural world even when we had not the slightest idea of the long-range consequences of what we were doing.[15]

Berry refers to the millennial vision that is found in the scriptural book of Revelation. It points to a time when all of life would be transformed and the human condition finally healed; we would experience peace, justice, and abundance. We became impatient for the divine to send us this new Earth, so we committed ourselves to bringing it into being ourselves. But our drive to bring this time of bliss into being has not eliminated our inner sense of loss and dissatisfaction.[16] This attitude toward the natural world confirmed a discontinuity between the human and the nonhuman that could easily justify exploitation.[17]

In contrast, classical civilizations and indigenous traditions lived within a very different sense of the world, one in which the human, the divine, and the cosmos were present to each other in an intimate way. Peoples understood that all of the forces of the universe and all realms of being would

14. See Berry, *The Sacred Universe*, 136: "Until recently, there has been a feeling in most religious traditions that spiritual persons were not concerned with any detailed understanding of the biological order of Earth. Often, the spiritual person was in some manner abstracted from concern with the physical order of reality in favor of the interior life of the soul. If attention was given to the physical order, this was generally in the service of the inner world."

15. Berry, *The Christian Future*, 40. Berry notes that the Christian affirmation of incarnation requires that we develop a greater sense that humans are part of the Earth community: "If God has desired to become a member of [the Earth] community, humans themselves should be willing to accept their status as members of that same Earth community" (ibid., 11). See also ibid., 33.

16. See Berry, *The Sacred Universe*, 13: "Ignorant of any spiritual significance in what we are doing, we remain profoundly dissatisfied, inwardly starved, spiritually and humanly debilitated, and unable to carry out successfully our finest endeavors."

17. See ibid., 138: "I propose that one of the most fundamental sources of our pathology is our adherence to a discontinuity between the nonhuman and the human, which gives all the inherent values and all the controlling rights to the human."

help humans to connect with their deepest selves, to overcome their fears and anxieties about the human condition, and to find the direction they need to take to participate in the full meaning of life.

But Berry says that in our contemporary world, human persons do not truly live in a universe. We tend to live in cities and countries, in economic systems, in cultural and perhaps religious traditions. Our alienation from the natural world is so extensive that we are not even aware of it. Even the idea that we should have an integral and intimate relationship with the natural world lies so far outside our horizons that we cannot contemplate it. Berry describes this cultural pathology or deep alienation in this way:

> While we have more scientific knowledge of the universe than any people ever had, it is not the type of knowledge that leads to an intimate presence within a meaningful universe . . . .Our world of human meaning is no longer coordinated with the meaning of our surroundings . . . . Our children no longer learn how to read the great Book of Nature from their own direct experience. They seldom learn where their water comes from or where it goes. We no longer coordinate our human celebrations with the great liturgy of the heavens . . . .We no longer hear the voice of the rivers, the mountains, or the sea . . . .The world about us has become an "it" rather than a "thou."[18]

As a consequence of this alienation, we live our daily lives in a world of objects, not subjects. We have little contact with the natural world; we regard it as a backdrop to our human undertakings. Indeed, it has little connection for us with what is meaningful in life. This is precisely why Berry articulates the need for a new story of the emergence of the universe that could provide humans with a new orientation and perspective, a context for connection, for purpose, meaning, and action. Through this story, Berry is seeking a comprehensive foundation that will nurture an intimate relationship between humans and the other-than-human world. However, we need to know that the story of the universe is a sacred story. As such, the story becomes a new cosmology that can help us, in our time, to truly know our place in the emerging universe.

A more adequate ecological spirituality is emerging out of this new story of the universe and the comprehensive consciousness that realizes there is a single community of Earth: "we form a single sacred society with

18. Berry, *The Great Work*, 15, 17. See also: "Everyone lives in a universe; but seldom do we have any real sense of living in a world of sunshine by day and under the stars at night. Seldom do we listen to the wind or feel the refreshing rain except as inconveniences to escape from as quickly as possible" (ibid., 54).

every other member of the Earth community, with the mountains and rivers, valleys and grasslands, and with all the creatures that move over the land or fly through the heavens or swim through the sea."[19] We begin to be able to hear the voices of all creatures. We learn of the integral relationship among all the members of the community of Earth and of the universe. Nothing is what it is without everything else. Everything, every component member, has its own identity, dignity, and inner spontaneity—this is its sacred dimension. Within this horizon, we can find ourselves relating as subjects to subjects, no longer alienated from each other, but living in a relationship of communion with all. If we could truly understand that our human story is integral with the story of the universe, "then we can see that this story of the universe is in a special manner our sacred story, a story that reveals the divine particularly to ourselves, in our times; it is the singular story that illumines every aspect of our lives—our religious and spiritual lives as well as our economic and imaginative lives."[20]

The opportunity to walk and contemplate the sacred story of the universe and to participate in ecology retreats are examples of ways in which we can experience a deep intimacy with the Earth, become present to the community of life, and truly know the universe story as our own story.[21] They are an embodiment of the cosmological spirituality that Berry says must become the context for our contemporary spiritual journeys: "We need to establish rituals for celebrating these transformation moments that have enabled the universe and the planet Earth to develop over these past many years . . . . To celebrate these occasions would renew our sense of the sacred character of the universe and planet Earth."[22]

This new story of the universe makes it clear to us that our sense of the divine is intimately related to the natural world. Berry states, "Why do we have such a wonderful idea of God? Because we live in such a gorgeous

19. Berry, *The Sacred Universe*, 85.

20. Ibid.

21. Between 1999 and 2012, I had the opportunity to participate in eight-day ecology retreats at Ignatius Jesuit Centre, Guelph, retreats rooted in the Spiritual Exercises of St. Ignatius of Loyola. Each day invited retreatants to reflect on a particular theme: what the voices of the community of life at the Centre has to say about God's love for the whole universe and each being; reflecting on the impact of our participation in ecological sin; noticing how the divine is revealed within creation; pondering what it means that God is intimately present in the passion and suffering of the Earth; celebrating the hope of the resurrection of the whole of creation; contemplating how we are to respond in love and live our lives as members of an integral, comprehensive, and sacred community. See https://ignatiusguelph.ca/ignatius-land/sacred-space/ and http://www.ignatianspirituality.com/ignatian-prayer/the-spiritual-exercises.

22. Berry, *Evening Thoughts*, 21.

world." Or again, "We would have no sense of the divine without creation. We seem not to realize that as the outer world becomes damaged, our sense of the divine is degraded in a corresponding manner."[23] What we are doing to the natural world affects our capacity to have a sense of the divine revealed in the natural world. If we could recover the depths of the meaning of the universe, then we could have greater intimacy with the divine that is manifested in the natural world. We would also have the type of foundation that we need for the transition into the Ecozoic era.

Knowing the universe as sacred, and knowing that our story and the stories of all other members of the Earth community are inseparable from the story of the universe, would help us to appreciate that all share this unity of origin. Berry points out that this means that everything that exists in the universe is genetically related to everything else. It also means that community is at the heart of the nature of existence, that there is a relationship of kinship of each being to every other being. "There is literally one family, one bonding, in the universe, because everything is descended from the same source . . . . On the planet earth . . . [w]e are literally born as a community; the trees, the birds, and all living creatures are bonded together in a single community of life."[24] To the degree that human beings can come to understand how intimate we are with the universe, the difficult, often impenetrable, psychic barrier between humans and the natural world can be removed. We can find ourselves relating as subjects to subjects, no longer alienated from each other, but living in a relationship of communion with all.

The Practice of Attentiveness and Intimacy

Contemporary nature writers can be important guides for us as we take the steps necessary for this conversion. I think, for example, of the work of Lyanda Lynn Haupt. In the beginning of her book, *Crow Planet: Essential Wisdom for the Urban Wilderness*, she says:

> There is a way to face the current ecological crisis with our eyes open, with stringent scientific knowledge, with honest sorrow over the state of life on earth, with spiritual insight, and with practical commitment. Finding such a way is more essential now than it has ever been in the history of the human species. . . . Our actions can rise instead from a sense of rootedness,

23. Berry and Clarke, *Befriending the Earth*, 8.

24. Ibid., 14–15.

> connectedness, creativity, and delight. But how are we to attain such intimacy, living at a remove from "nature," as most of us do, in our urban and suburban homes?[25]

Haupt reminds us that the reality is that we are not removed from "nature." We tend to think that we connect to nature only when we participate in a wilderness experience of some kind. The problem with this thinking is that we perpetuate a deep separation between our day-to-day lives, which we think have nothing to do with nature, and the wild places, which we regard as "true nature."

Haupt helps us to understand that the truth of the matter is quite different. We are connected to the natural world exactly in and through our everyday lives. "[I]t is in our everyday lives, in our everyday homes, that we eat, consume energy, run the faucet, compost, flush, learn, and live. It is here, in our lives, that we must come to know our essential connection to the wilder earth, because it is here, in the activity of our daily lives, that we most surely affect this earth, for good or for ill."[26]

This truth that we are connected to the natural world in and through our everyday lives, that our everyday lives are part of an emerging, evolving cosmic story, requires us to start walking the paths of our neighborhoods, to start knowing the breadth of all of our neighbors, human and nonhuman, "on and off the concrete, above and below the soil."[27] We must begin to pay attention to the places where we live, to all the members of the community of life where we live, and to understand our intricate net of connections with the rest of the earth community. Such attention is the only way to cultivate the types of insights into the natural world that we need, insights that are based in attention, knowledge, and intimacy. For Haupt, as for Berry, "an intimate awareness of the continuity between our lives and the rest of life is the only thing that will truly conserve the earth—this wonderful earth that we rightly love."[28]

Haupt points to Aldo Leopold's challenge to us in *A Sand County Almanac*: "The reckoning Leopold asks of us requires the cultivation of insight based in attention, knowledge, and intimacy. It asks that we pay loving attention to the places we live, to understand their intricate net of connections with the wider earth."[29] She encourages us to develop the qualities of an urban naturalist: studying our local ecosystems; learning the names of

25. Haupt, *Crow Planet*, 7–8.
26. Ibid, 9.
27. Ibid, 13.
28. Ibid, 12.
29. Ibid, 8.

creatures; learning to respect the wildness of all animals; working with a particular question;[30] carrying a notebook with you and maintaining a "field trip" mentality; making time for solitude; knowing that we stand in a long line of people who have stressed the value of personal observation and attention to the places where we live.

Barbara Brown Taylor states that the practice of paying attention, especially to the natural world, teaches us reverence: "The easiest practice of reverence I know is simply to sit down somewhere outside, preferably near a body of water, and pay attention for at least twenty minutes . . . . With any luck, you will soon begin to see the souls in pebbles, ants, small mounds of moss, and the acorn on its way to becoming an oak tree."[31] Thomas Lowe Fleischner, editor of *The Way of Natural History*, says that "'natural history' is a practice of intentional, focused attentiveness and receptivity to the more-than-human world . . . . Attention is prerequisite to intimacy. Natural history, then, is a means of becoming intimate with the . . . world."[32] Fleischner goes on to argue that attentiveness to nature matters because,

> in a very fundamental sense, we are what we pay attention to. . . . Our attention is precious, and what we choose to focus it on has enormous consequences. What we choose to look at, and to listen to—these choices change the world.[33]

As we walk and develop our capacity to pay attention to the natural world, as we become more intimate and present to the members of the community of life, we can begin to know that our human story is integral with the story of the universe.

Conclusion

In 2015, Pope Francis issued the encyclical letter *Laudato Si'*, the first major document of the Catholic Church that deals, in its entirety, with the ecological crisis. At its heart is the question: "What kind of world do we

30. Haupt quotes E. O. Wilson: "You start by loving a subject. Birds, probability theory, stars, differential equations, storm front, sign language, swallowtail butterflies. . . . The subject will be your lodestar and give sanctuary in the shifting mental universe." *Crow Planet*, 57

31. Taylor, *An Altar in the World*, 22–23. Taylor explores a number of simple practices that can help us to discover the sacred in the small things we do and see in our daily lives.

32. Fleischner, ed., *The Way of Natural History*, 5.

33. Ibid, 9.

want to leave to those who come after us, to children who are now growing up?"[34] This question, and this document, reflects Thomas Berry's lifelong concern for all of the Earth's children; human and other-than-human alike. Francis's presentation of the Judeo-Christian witness leads him to speak of a central theme, namely, that everything is interconnected, that we are part of a universal communion.[35] Like Berry, Francis has hope that human beings can undergo ecological conversion and find new, more appropriate ways of dealing with the devastation of our common home.[36]

The notion of an integral ecology is key to the encyclical's proposals for responding to the ecological crisis. With echoes of Berry's call for an understanding of the human as an integral member of the Earth community, Francis has in mind a new paradigm of justice, and a perspective in which "[n]ature cannot be regarded as something separate from ourselves or as a mere setting in which we live." Instead, we need to understand that "we are part of nature, included in nature, in constant interaction with it."[37]

This development of an ecological culture (or, in Berry's words, the Ecozoic Era), is helped by an ecological spirituality. And the guide that Pope Francis turns to, as do so many other writers, is St. Francis of Assisi, "the example par excellence of care for the vulnerable and of an integral ecology that is lived out joyfully and authentically."[38] "Just as happens when we fall in love with someone, whenever Francis would gaze at the sun, the moon or the smallest of animals, he burst into song."[39] What St. Francis offers us is a kinship model of relationship to the rest of the Earth community. It starts from the place that we human beings, all of us, are part of creation. We are not aliens on this Earth; we are made of the same elements, part of the same family, sharing the same DNA with most of the rest of creation. What St. Francis models for us is how to feel intimately united with all that exists. As Pope Francis says in *Evangelii Gaudium* and repeats in *Laudato Si'*, "God has joined us so closely to the world around us that we can feel the

34. Francis, *Laudato Si'*, n. 13.

35. "All of us are linked by unseen bonds and together form a kind of universal family, a sublime communion which fills us with a sacred, affectionate and humble respect." (Ibid, n. 89).

36. See Ormerod and Vanin, "Ecological Conversion."

37. Francis, *Laudato Si'*, n. 139.

38. Ibid, n. 10. St. Francis is the patron saint of those who work on ecological issues.

39. Ibid, n.11. See Gagnon et al., "You Love All That Exists . . . . All Things Are Yours, God, Lover of Life." This letter makes clear that we are called to live out of the horizon of ultimate, divine loving, which is a loving of the whole cosmos, and of each individual being in that cosmos.

desertification of the soil almost as a physical ailment, and the extinction of a species as a painful disfigurement."[40]

This is the depth of intimacy that an ecological spirituality can nurture in us. It is the kind of intimacy that we can learn from contemporary nature writers about how to be attentive to the community of life within which we live, how to understand the nature, role, value, and dignity of all members of the community, how to affirm the whole of the cosmos as the most comprehensive context of our being. This is the foundation of a cosmological spirituality that changes our hearts and minds, making it possible for us to become, with God, knowers, co-healers, and lovers of all that exists.

Bibliography

Berry, Thomas, and Thomas Clarke. *Befriending the Earth: A Theology of Reconciliation Between Humans and the Earth.* New London, CT: Twenty Third, 1991.

Berry, Thomas. *The Sacred Universe: Earth, Spirituality, and Religion in the Twenty-First Century,* edited by Mary Evelyn Tucker. New York: Columbia University Press, 2009.

———. *Evening Thoughts: Reflecting on Earth as Sacred Community.* Edited by Mary Evelyn Tucker. San Francisco: Sierra Club, 2006.

———. *The Christian Future and the Fate of Earth.* Edited by Mary Evelyn Tucker and John Grim. Maryknoll, NY: Orbis, 2009.

———. *The Great Work: Our Way Into the Future.* New York: Bell Tower, 1999.

Fleischner, Thomas Lowe, ed. *The Way of Natural History.* San Antonio, TX: Trinity University Press, 2011.

Francis. Apostolic Exhortation *Evangelii Gaudium.* United States Conference of Catholic Bishops, November, 2013.

———. *Laudato Si'.* Rome: Liberia Editrice Vaticana, 2015.

Gagnon, Jean, et al. "You Love All That Exists . . . . All Things Are Yours, God, Lover of Life." Pastoral Letter on the Christian Ecological Imperative, Canadian Conference of Catholic Bishops, Social Affairs Commission, October, 2003.

Haupt, Lyanda Lynn. *Crow Planet: Essential Wisdom from the Urban Wilderness.* New York: Back Bay, 2009.

Lonergan, Bernard. *Method in Theology.* New York: Herder and Herder, 1972.

Ormerod, Neil, and Cristina Vanin. "Ecological Conversion: What Does It Mean?" *Theological Studies* 77.2 (2016) 328–52.

Swimme, Brian, and Thomas Berry. *The Universe Story: From the Primordial Flaring Forth to the Ecozoic Era—A Celebration of the Unfolding of the Cosmos.* New York: HarperSanFrancisco, 1992.

Taylor, Barbara Brown. *An Altar in the World: A Geography of Faith.* New York: HarperOne, 2009.

40. Ibid, n. 89; see Pope Francis Apostolic Exhortation *Evangelii Gaudium*, 315.

COSMOS AND EARTH

Encountering Earth from a Scientifically Informed Theological Perspective

—JAME SCHAEFER—

Fireflies piercing the darkness of my childhood back yard.
Grasshoppers leaping long distances on the grass.
A bright red cardinal feeding hungry hatchlings in a nest below my bedroom window.
Sap pouring from sugar maples in a neighbor's forest.
Pine seedlings emerging in a fire-blackened Yellowstone Park forest.
An enormous grizzly bear sauntering up a mountain path in Waterton Lake Park.
Petrified wood jutting from Specimen Ridge.
Coral reefs off the Belizean coast teeming with colorful fish.
Miles of wildflowers adorning Paintbrush Canyon in the Tetons.
Blue footed boobies[1] dancing in the Galapagos.
Hundreds of butterflies[2] swarming over my head while resting in the conical depression atop Mt. Lola in the Sierra Nevadas.
Clear blue skies dotted with luminous cloud formations.
Ever-changing colors of Lake Michigan.
Awesome!

1. Blue Footed Booby Mating Dance, November 1, 2013; example available at https://www.youtube.com/watch?v=oYmzdvMoUUA.

2. Northwestern Fritillary butterfly, *Speyeria hesperis irene*; image available at http://www.pbase.com/tmurray74/image/63302284.

As a child, I wondered why fireflies beamed, grasshoppers could jump so far, a pair of cardinals shared duties so precisely and resourcefully when feeding their hatchlings, corals attracted a plethora of other marine species, trees petrified, and birds danced so cleverly. Some answers were relatively easy to find when paying closer attention and/or consulting sources of information. Why the magnificent bear did not notice trembling hikers curled around uphill Lodgepole pines became plausible when passing a vast area of denuded blueberry bushes upon our descent. Connections among animals and plants in their shared habitats awakened my longing to know more about their interdependence.

Feelings of disgust loomed as I encountered blights on the awesome and wondrous. Disgust when the thick stream of yellow-brown emissions from an electricity plant fueled by high-sulfur coal cut through the blue sky. Disgust when an interstate highway was proposed to be built through a forest of sugar maple trees in Manitowoc County. Disgust when passing over a blackened area in the middle of lush green where a mountain top in West Virginia had been severed to remove coal. Disgust when more nuclear power plants were proposed to be built north of where I lived and throughout the United States despite the federal government's failure to isolate the highly radioactive used fuel from the biosphere. Disgust when finding recyclable plastics thoughtlessly trashed with non-recyclables in the community room of Marquette Hall. Disgust when watching YouTube clips showing vast islands of floating plastics in the Pacific, Atlantic, and Indian oceans. Disgust when seeing pictures of more elephants brutally slaughtered for their ivory in Kenya, the Democratic Republic of Congo, and the Central African Republic. Disgust when high concentrations of polychlorinated biphenyls (PCBs) and heavy metals from industrial processes were discovered on the banks and in the sediments of the Sheboygan River and Harbor. My visceral reactions prompted a desire to know more about these problems, identify others in the community who shared my concerns, and take action to address them.

Thus began my cognitive engagement in issues that threaten humans, other species, ecological systems, and the biosphere. I share this journey in two sections. The first focuses on key problems I addressed in various secular capacities that eventually prompted my interest in academically exploring religiously grounded motivation for acting. In the second section, I discuss my encounters with Earth as a theology professor, researcher, writer, and collaborator with other specialists in advancing the study of ecological ethics.

Encountering An Imperilled Planet—Key Problems

The US Department of Transportation's proposal to pave an interstate highway through prime agricultural land along the Lake Michigan shoreline, the electric utilities' plan to locate a nuclear power plant along the Lake Michigan shoreline, and the discovery of PCB concentrations in the Sheboygan River and Harbor surfaced sequentially in Manitowoc and Sheboygan counties late in the 1970s. All occasioned opportunities for me to encounter Earth. Threats to the land, air, and water stimulated responses by small farmers whose land was slated to be taken according to the law of eminent domain, by charter boat captains and fishermen who could no longer boast about the quality of fish to be caught, and by a variety of people who lived in proximity to the proposed nuclear plant site. Many began to organize and act. Prominent among the activists were people from all walks of life who expressed their concerns from various moral, spiritual, and religious perspectives and who sought scientific and technical information to support their arguments.

Interstate 43

Small farmers sparked action on the proposed interstate corridor. One was Ed Klessig, who had taken a course from Aldo Leopold at the University of Wisconsin in Madison and was fired up by his teacher's famous "land ethic."[3] The interstate was slated to cut into part of the Klessig family's treasured maple forest, where my family collected sap each February. We carried buckets to the trays where the sap was boiled into the most delicious maple syrup and celebrated with pancakes doused with butter made from the milk of the homestead's cows and, of course, the maple syrup. Those of us who participated in this annual tradition appreciated the maple forest, the cows, and the rich soil joined them and other farmers to plot a course of action.

Planning to protest required knowledge about the interstate system, the process through which a proposal is considered and approved, and junctures at which to act effectively. Arguing against locating the highway through prime agricultural land also required knowledge of the taxonomy of the rich soil that formed over hundreds of years to yield fertile ground for growing staple crops. As support for the interstate highway welled among construction workers, owners of businesses, and local governments who anticipated profiting from the highway, we countered their arguments and cautioned businessmen in the area against believing that they would gain

3. Leopold, *A Sand County Almanac, and Sketches Here and There*, 224–25.

any benefit as the highway bypassed their communities. We also underscored the value of the soil and the forests that could not be monetarily quantified. The pros and cons were aired in public fora, during hearings held by the Wisconsin and US departments of transportation, and in an in-depth study by the League of Women Voters (in which I participated).

To attract attention about the loss of prime agricultural land and the maple forests through which the interstate would cut, about 600 people carrying signs and banners marched, rode bikes, and hauled children in wagons and baby carriages along a highway that paralleled the proposed corridor. To reflect as a community, we celebrated a special Catholic liturgy[4] in the front yard of the Francis and Nancy Salm Homestead, during which we expressed gratitude to God for the rich land and forests and the families whose homesteads would be adversely affected. Late one evening, the Klessigs loaded eight of their herd onto trailers, drove 125 miles to the state capital, and camped in front of the governor of Wisconsin's office for a week. Local to national media covered the campout, the farmers who eloquently expressed their deep appreciation for the land and forests, and the people who purchased cow milk and manure.

Despite these efforts, the construction of Interstate 43 was approved and completed in 1981. The only positive aftermath of this otherwise sad experience were the deep friendships established with creative, caring, and committed environmentalists who remain my best friends today. Though I had not anticipated the eventual effect on my thinking at that time, I began to yearn for the ability to theologically express my appreciation for the land, forests, and people who cared about and for them.

The Proposed Haven Nuclear Power Plant

While protest walking part of the corridor slated for the new interstate highway, some teachers in the Sheboygan and Catholic school systems continued conversations we had begun earlier about the proposal to build a nuclear power plant in the northernmost part of Sheboygan County. The land had been used for anti-aircraft testing from 1949 to 1959 but sold subsequently to Wisconsin Power and Light Company. We began a study group, volunteered to search for information on particular aspects of the proposed plan (e.g., safety of pressurized water reactors, emissions of low-level radiation,

4. Identifying appropriate scriptures for reading at the mass, writing prayers of petition, and preparing offertory gifts proved to be a stimulating experience for thinking deeply about the meaning of scriptures and writing prayerful petitions that were applicable to the soil, trees, animals, and farmers.

storage and disposition of the highly radioactive used fuel that would have to be removed from the reactors, and the status of nuclear plant decommissioning technology), the need for another nuclear power plant, and alternatives for meeting electricity needs (e.g., energy efficiency and renewable sources).

Volunteering for the used fuel category, I combed every possible source of information I could find from the US Nuclear Regulatory Commission, the US Energy Research and Development Administration,[5] the US Environmental Protection Agency, the nuclear industry, scientific journals, and environmental organizations.[6] I concluded that no more nuclear power plants should be approved for construction at Haven or anywhere else in our country until the federal government began isolating the hazardous used fuel removed from the nuclear reactors. Future generations should not be saddled with managing and trying to figure out a way of isolating from the biosphere this lethal byproduct of electricity generation. Because the bundles of used fuel removed from the reactors remained in temporary storage in concrete pools at the plants many years beyond the utilities' expectations, the pools had to be re-racked for more dense storage. Some owners of nuclear plants had begun to move some of the older bundles outside where the radioactivity would continue to decay and heat would dissipate in the ambient air. Despite promises by the federal government that the used fuel would be removed for disposal, promises that the utilities repeated frequently when proffering the construction of the Haven plant, its removal and disposition was not assured. Approving the construction of yet another nuclear facility in the absence of a system for isolating the used fuel from the biosphere was both imprudent and intergenerationally unjust.[7]

After sharing our conclusions on issues each of us researched, we decided that the only justifiable course of action was to oppose the construction of the proposed Haven nuclear plant. A plan of action was developed, and Safe Haven, Ltd. was launched. People from all walks of life and occupations participated in educational programs, discussions with electric utilities' executives and government officials, peaceful demonstrations, media events, petitions to local governments, religious services, an energy efficiency fair, and a tour of a local solar-powered home. I led or participated in all of these efforts. One that was particularly poignant was the consecration

5. Subsequently renamed the Department of Energy.

6. A challenge at that time prior to the ease with which information can be sought over the Internet today.

7. A brief history of the federal government's failure to provide a means for isolating the spent fuel from the biosphere is included in Schaefer, "Imprudence and Intergenerational Injustice."

of the 600-acre Haven site by a Catholic priest as the Children's Peace Park, after which a rabbi and a minister shared their faith-based concerns about the proposed nuclear plant.[8] Teachers, farmers, shopkeepers, engineers, homemakers, scientists, students, physicians, lawyers, and religious leaders testified at hearings held by the Nuclear Regulatory Commission in the Sheboygan County Court House.[9] Officials of the local Chamber of Commerce, businessmen who hoped to gain from selling concrete and other materials that would be used to construct the facility, and local governmental officials who anticipated tax benefits voiced their support for the proposal. The Wisconsin utilities weighed their options while members of Safe Haven continued their educational and advocacy efforts. By 1980, the utilities cited economic reasons for withdrawing their application before the Nuclear Regulatory Commission to construct two reactors on the Haven site.[10]

Efforts to address the nuclear used fuel disposition issue continued in other venues. Before the utilities withdrew their application for the Haven plant, the Public Service Commission of Wisconsin initiated its first adjudicatory proceeding on the utilities' advance plans for meeting the state's electricity needs, and a simple letter of inquiry about this formal proceeding resulted in my representing Safe Haven as a party. Bolstered by questions and corresponding answers that engineers and scientists helped develop for cross-examining the utilities' and Public Service Commission's witnesses and inspired spiritually by members of religious communities who silently prayed while I played lawyer, we helped build a lengthy record of the proceeding. The commissioners voted unanimously in 1978 to bar planning to add more nuclear capacity in Wisconsin until the uncertainties pertaining to spent fuel disposition, plant decommissioning, and uranium fuel availability are resolved.[11]

Another effort before the Nuclear Regulatory Commission in its first Waste Confidence Rulemaking Proceeding did not end as positively. Though

8. "Haven Site Blessed as 'Peace Park,'" *The Milwaukee Journal*, June 3, 1979.

9. Safe Haven sponsored workshops facilitated understanding the hearing process and preparing testimonies. These workshops became parties! The president of Safe Haven, Bill Hanley, who was well versed in Saul Alinsky's tactics, insisted that we have fun amidst the gravity and demands of our objectives. Hanley's consistent mantra, "Remember folks, they have the electricity, but we have the power" emboldened opponents of the proposed Haven facility.

10. One reactor was cancelled in 1978 and the other in 1980.

11. Public Service Commission of Wisconsin, Advance Plan. A state law was passed in 1983 declaring a moratorium on building new nuclear power plants in Wisconsin until a spent fuel disposal system was available that assured containment of all the spent fuel from nuclear power plants in the state. That moratorium was lifted in 2016 by a bill that Governor Scott Walker signed.

I appeared at the oral arguments held by the Commission in Washington, DC, presented petitions signed by over a thousand people who lived in the Sheboygan and Manitowoc areas, brought reams of oral tapes on which their no-confidence testimonies were recorded, and submitted statements signed by local governments and organizations indicating widespread lack of confidence in the federal government to isolate the highly radioactive used fuel from the biosphere, the five commissioners voted their confidence in 1983 that "one or more mined geologic repositories for commercial high-level radioactive waste and spent fuel will be available by the years 2007–9."[12] Obviously, the commissioners' confidence had been highly misplaced. A geologic repository nor any other system for isolating the used fuel does not exist more than sixty years after nuclear fuel began to be used to generate electricity in the United States.

Before the Nuclear Regulatory Commission had rendered its decision, the Governor of the State of Wisconsin asked me to serve on the newly established Radioactive Waste Review Board that was tasked with responding to the US Department of Energy's interest in studying two granite formations within the state for their capacity to isolate used fuel from the biosphere.[13] I served as its first chairperson during a tumultuous period of time that is chronicled in a paper published by the Center for Public Affairs at the University of Wisconsin-Green Bay in 1988.[14] My primary effort during the five years I served aimed at demanding a scientifically and technically sound process of repository siting that was thoroughly transparent, culturally sensitive, and open to involving the states, local governments, and Native Americans. US senators and congressmen who represented the seventeen states with sizeable granite deposits identified by the Department of Energy fiercely opposed the project. In 1986, the Department of Energy cancelled the search for a repository site in crystalline rock and turned its full attention to the volcanic tuff in Yucca Mountain, Nevada—a state represented in the House of Representatives at that time by only one Nevadan. President Obama cancelled the project in 2009 as he had promised during

12. US Nuclear Regulatory Commission, "Rulemaking on the Storage and Disposal of Nuclear Waste."

13. Having already served in Governor Lee Dreyfus's Energy Task Force from 1980–981, he and his staff were aware of my opposition to more nuclear capacity in the State of Wisconsin primarily due to the used fuel disposition failure.

14. Schaefer, "State Opposition to Federal Nuclear Waste Repository Siting: A Case Study of Wisconsin 1976–1988." Based on lessons learned from the Wisconsin experience, I included in this document some recommendations for proceeding to resolve the spent fuel disposition impasse. An earlier synopsis of this case study was included in Schaefer, Kraft, and Clary, "Politics, Planning, and Technological Risk."

his campaign, though he insisted subsequently that more nuclear generated electricity should be part of the US energy mix.[15]

Toxicants in the Sheboygan River and Harbor

A third major problem that warranted attention and action was the discovery in 1977 of high concentrations of PCBs, heavy metals, and other toxicants in the Sheboygan River and Harbor. They also had been discovered in other areas around the Great Lakes and attributed primarily to industrial activities in their locales. PCBs were particularly problematic to human health, scientific researchers concluded, because they bioaccumulate in body cells as they move up the food chain and present a carcinogenic risk to human health when contaminated fish and wildlife are consumed.[16]

Fish advisories were issued and posted along the River and Harbor. Because increasing numbers of Hmong people had moved into the area where they fished for the protein staple in their diets, advisories were translated into their language and workshops were held telling them which fish to avoid and how to cut the fatty tissue from the fish they caught before cooking them. This service was particularly important to those of us who served as sponsors of Hmong families that had fled Laos when the Communists assumed control of their country.

For a community that relished its Friday fish fries, charter boat captains who supported their families through day-long fishing expeditions, public health specialists who expressed concern about contamination of Lake Michigan which supplied water to the 58,000 member community, physicians who were alert to studies identifying the effects of PCBs on the brains of unborn babies whose mothers consumed contaminated fish and babies whose mothers nursed them, and environmentalists who were aware of impairments on fish and waterfowl, this problem had to be addressed. The Wisconsin Department of Natural Resources and the US Environmental Protection Agency worked with local governments and residents on steps that had to be taken to remediate the area. In 1986, the Environmental Protection Agency designated the lower fourteen miles of the Sheboygan River as a Superfund Site due public health threats.[17] The following year, the International Joint Commission prioritized the Sheboygan River and

15. Office of the Press Secretary, "FACT SHEET." More details are available in Schaefer, "Imprudence and Intergenerational Injustice."

16. Agency for Toxic Substances and Disease Registry, "Toxic Substances Portal."

17. U.S. Environmental Protection Agency, "Site Information for SHEBOYGAN HARBOR & RIVER."

Harbor with forty-one other contaminated sites in the Great Lakes Basin as "areas of concern."[18] A remedial action plan for the River and Harbor was developed and initiated,[19] and, as lawyers haggled over their clients' financial responsibilities, sediments containing high concentrations of PCBs were removed by 2012 with financial support from the Great Lakes Restoration Initiative.[20] Some remediation targets were met while others remain at various stages of completion. Monitoring is ongoing by the Wisconsin Department of Natural Resources and the US Environmental Protection Agency.[21]

Throughout the early part of this process, I served as the local administrator of a grant awarded to the Lake Michigan Federation by the Environmental Protection Agency for organizing public information sessions and facilitating input in remedial action planning. One especially informative event was a canoe trip with scientists who pointed to and discussed areas where sediments had to be dredged, soil along the shore that had to be scraped, and experiments in treating PCBs were underway. The stretch of the river through which we canoed was lovely with hanging trees, lush foliage, and fish but marred by fenced-off areas to deter people from contact with the contaminated soil that was slated for removal.[22]

My most formative experience on this issue resulted in a stumble into academia. That occurred after the Science Advisory Board of the International Joint Commission asked me to suggest a way of substantively involving the public in addressing the persistent toxicants and other nagging problems in the Great Lakes Basin that were affecting the quality of water and, thereby, the quality of life. After considerable weighing of possibilities, I prepared a discussion paper in which I urged the Science Advisory Board to propose the initiation of a project aimed at developing a code of ethics

18. The International Joint Commission was established in 1909 by the United States and Canada to cooperate in safeguarding their boundary waters. An overview of the forty-three areas of concern with links to pertinent documents is provided by Day, "International Joint Commission Releases Special Report on Pollution Cleanup Efforts in Great Lakes Areas of Concern.

19. Department of Natural Resources, "The Sheboygan River Remedial Action Plan," 1989 and 2014.

20. US Environmental Protection Agency, "Great Lakes Restoration Initiative (GLRI)."

21. Office of the Great Lakes, "2015 Remedial Action Plan Update for the Sheboygan River Area of Concern."

22. Understanding the risks of contact with PCBs and other toxicants proved particularly helpful when I was asked by the State of Wisconsin's Division of Health in 1988 to coach epidemiologists on constructive ways of engaging concerned people who lived near heavily contaminated sites.

for functioning in the Great Lakes Basin, provided a list of philosophical sources that could serve as a basis for conversation, and identified categories of groups to engage in this project—local governments, Native American tribes, nongovernment organizations, and communities of various types.[23] My subsequent discussion with members of the Science Advisory Board led to their request to submit a follow-up report listing specific organizations in the Great Lakes Basin to draft the code of ethics and details for a step-by-step process through which it would be developed. As I prepared this report, I realized the importance of inviting religious communities whose members might be motivated by their faith to think about how they should function.[24] I also realized that I needed to delve more deeply into the scriptural and theological sources of my own religion to see what I could find that might be motivating for Catholics. By the time the Commission discussed and tabled its Science Advisory Board's proposal to initiate the creation of a code of ethics for functioning in the Great Lakes Basin, I was immersed in religious studies at the doctoral level—the next stage of my encounter with Earth.

Encountering Earth as an Academician

Yearning to move beyond my independent forays of Catholic theological sources, I decided to enroll in a graduate level ethics course at Marquette University. That decision launched me into encountering Earth on a cognitive level that led to teaching, researching, collaborating with colleagues, and, of course, publishing. Throughout, I was buoyed by direct experiences with Earth when hiking in the mountains, snorkeling above coral reefs, and amazing over the ever-changing hues of Lake Michigan.

Studying at the Graduate Level

When I began the ethics seminar at Marquette with the hope of becoming more informed about the Catholic theological tradition and its potential for relating to ecological concerns, the Jesuit professor alerted me to his unfamiliarity with the religion-science field but assured me that I could read and reflect on assigned sources, select research topics, and write required papers through an ecological lens. I found possibilities in some sources and

23. Schaefer, "Toward a Code of Ethics for the Great Lakes Basin Ecosystem."

24. Schaefer, "Creating a Code of Ethics for the Ecosystem."

yearned for more. One course led to another, at which point I was required to apply for the doctoral program or discontinue. I was encountering Earth from a theological perspective informed by the natural sciences, and my reflections pointed to trajectories in behavior that I found promising for ecological concerns of the type I had already addressed politically, economically, and morally. My theological reflections provided a deeper, faith-based way of encountering Earth, her ecological systems, other species, and their habitats. I found the freedom to read, think, and engage in theological discourse stimulating, hopeful, and motivational. The greatest inspiration came from eminent theologians during the patristic and medieval periods, especially Thomas Aquinas. My doctoral qualifying exams brought me to the height of my academic acumen, but I could bask in their laurels for only a few moments as I prepared to testify before the US Nuclear Regulatory Commission on the proposed re-racking of spent fuel storage at the Kewaunee Nuclear Power Plant in Manitowoc Country.

Proceeding to my dissertation topic, I settled on Aquinas's ideas about the unity and diversity of creation and his systematic treatment of the moral virtues. I placed my exploration and reflections within the context of the Great Lakes Basin. According to members of my family who were present at my defense, it was more like a celebration because my efforts were affirmed as Marquette's first dissertation relating theology and ecology.[25] I felt well equipped to return to advocacy work within the Archdiocese of Milwaukee and to continue to consult with governments on ecological problems while writing and speaking from a Catholic theological perspective.

Teaching

I had not planned to teach after receiving my doctoral degree in Religious Studies. However, when a Marquette professor become ill and I was asked to assume his courses, I agreed to serve. Almost immediately I found interacting with undergraduate students gratifying, especially when I was able to open them to relating constructively to the natural sciences. Midway through that first semester, I was asked to teach the Religion and Science course that was in jeopardy of being eliminated because it had not been offered for several years and no one among the faculty was interested in teaching it. I agreed to develop a syllabus for the course and decided to prepare another that focused on religious foundations for ecological ethics. Preparing for and teaching two sections of Religion and Science underscored for

25. Schaefer, "Ethical Implications of Applying Aquinas's Notions of the Unity and Diversity of Creation to Human Functioning in Ecosystems."

me the necessity of discourse about God and the human person informed by contemporary scientific findings. In their evaluations of the course, students indicated their feelings of "peace" and "relief" that they learned how to relate their religious faith and the natural sciences constructively and did not have to choose between the two. What more could spur a theologian to continue to teach?

Adding to this gratification was an award in 1996 from the Templeton Foundation for my Religion and Science course. A detailed description of it was included with four others in the promotion package sent to colleges and universities throughout the world to encourage teaching religion-science topics, and I was invited to participate in a week-long workshop at Oxford University featuring presentations by and consultation with major scientist-theologians.[26] Another award followed that provided funding for guest religion-science lectures and related activities at Marquette. When the chairperson of the Department of Physics approached the chairperson of my department to identify faculty who might help develop a symposium in which our disciplines would be related, I was tapped for this task. That request eventually resulted in the first offering of a team-taught course in Spring 1998—The Origin and Nature of the Universe.[27] During the next semester, colleagues in physics, philosophy, and I formed the Albertus Magnus Circle to encourage faculty discussion of issues at the boundaries of theology, philosophy, and the natural sciences. Our interactions helped us hone an approach to discourse that has been constructive and productive—defining our terms, realizing our assumptions, and avoiding the conflation, confusion, and compartmentalizing of our disciplines in our quest to engage in cogent and meaningful discourse about God, the world, and humanity.

As these religion-science endeavors were unfolding, I was allowed to offer the course on religious foundations for ecological ethics and began to discuss with faculty in other departments the possibility of developing an interdisciplinary environmental program at Marquette. A tenured colleague and I (a visiting assistant professor at that time) received a grant to pursue this possibility and organized a workshop for faculty from pertinent departments and colleges to create a proposal to establish the Interdisciplinary Minor in Environmental Ethics. After a lengthy, in-flux approval process, the minor was approved and students were able to declare it for the first time in Fall 2001. My course in Religious Foundations for Ecological Ethics received

26. Arthur Peacocke, John Polkinghorne, Fraser Watts, and Niels Gregersen, joined by philosopher Mary Midgley.

27. The sixth and final offering of this team-taught course with Physics Professor Emeritus John Karkheck occurred during Spring 2016. An article is forthcoming on why and how we taught the course variously and their outcomes.

a dedicated number and was designated as one of the courses required for the minor. A capstone seminar was created, during which students would integrate the knowledge and skills gained in the required courses to address an ecological problem from an ethical perspective.[28] Each capstone I have offered has been exciting, demanding, and unfolding as new ways of encountering Earth with my students. Many have graduated from Marquette well prepared for graduate studies in biology, environmentally sustainable engineering, and environmental law, positions in government agencies and non-governmental organizations, and teaching.

Researching and Writing

These positive teaching experiences compelled me to accept a tenure track position in systematic theology and ethics at Marquette with specializations in relating Catholic theology and the natural sciences and exploring religious foundations for ecological ethics. In my research and publications, I have focused on patristic and medieval sources in the Catholic theological tradition that attracted my interest throughout my doctoral studies, and I continue to find inspiring ideas about which to write as I read primary sources. They are filled with fruitful ways of thinking about God's creation, responding positively to other species and systems of Earth, and identifying behavior trajectories (e.g., valuing, appreciating, respecting, cooperating, living virtuously, and loving). I realized, however, that appropriating ideas generated in the fourth to fifteenth centuries and applying them to ecological problems in the twenty-first requires caution and care. To avoid misrepresenting and misconstruing theologians who inspired my thinking, I developed the "critical-creative method" for appropriating concepts in their works that initially stimulated my thinking (step one), recognizing their prescientific view of the world, the contexts of the times that propelled them to write, and the meaning they most likely intended to convey (step two), probing and, if necessary, reconstructing the meaning of the concept today in light of our current understanding of the world (step three), considering the relevance of the concept for addressing ecological degradation (step four), and identifying the behavior trajectory suggested (step five). This method has been demonstrated in several of my publications[29] and

28. Electricity production and use in the United States, human-forced climate change, and environmental injustice in Milwaukee are among the problems addressed in INEE capstone seminars. A list of topics with reports prepared by the students is available at http://www.inee.mu.edu/CapstoneSeminarProjects.shtml.

29. One major example is Schaefer, *Theological Foundations for Environmental*

used successfully by my students to address ecological problems they have researched.

I also have engaged Catholic social teachings by popes and other bishops[30] who expressed concern about the ecological crisis and urged the faithful to reflection and responsible action. Mentioned by Pope Paul VI in 1971 as an emerging concern,[31] the ecological crisis became the focus of Pope John Paul II's message on the 1990 World Day of Peace[32] that Pope Benedict XVI extended in his message on the 2010 World Day of Peace.[33] Though I have reflected on these papal statements and specific principles of Catholic social teachings in essays and articles,[34] the first encyclical dedicated to ecological concerns that Pope Francis issued in 2015 has warranted extensive attention in my research agenda as well as in teaching and guest lecturing. *Laudato sí, On Care for Our Common Home* is a phenomenal document in which the pope explores problems that have grave ramifications for vulnerable people, draws upon teachings by other bishops and reflections by theologians in the Catholic tradition, and calls all people to dialogue and action on caring for Earth.[35] Many reflections on his encyclical have cumulated to an array of works by theologians and other specialists. My most recent contribution is an examination of virtues that surface implicitly in the encyclical that I think constitute manifestations of ecologically conscious people who have accomplished the "ecological conversion" to which the pope calls everyone.[36] The body of literature generated by theologians and others on this epochal encyclical and the scientific evidence of the interconnections of humans and other species within systems of Earth warrant naming a more

Ethics.

30. Recall that the pope also serves as the Bishop of Rome.

31. Paul IV, *Octogesima Adveniens.*

32. John Paul II, *Peace with God the Creator, Peace with All of Creation.* This was the first papal statement dedicated to the ecological crisis; as a statement, it does not carry the significance of an encyclical—a major teaching by a pontiff that the faithful should seriously consider.

33. Benedict XVI, *If You Want to Cultivate Peace, Protect Creation.*

34. From 2002–2011, I convened groups on Catholic Theology and Ecology and on Catholic Theology and Global Warming at conventions of the Catholic Theological Society of America during which theologians reflected on their research. The latter yielded an anthology, Schaefer, ed., *Confronting the Climate Crisis*, in which I explored implications of extending key principles of Catholic Social Teaching to ecological concerns. Other efforts include Schaefer, "Environmental Degradation, Social Sin, and the Common Good."

35. Francis, *Laudato sí.*

36. Schaefer, "Converting to and Nurturing an Ecological Consciousness –Individually, Collectively, Actively."

inclusive category than "social." I am in the process of filling that lacuna by researching major principles of Catholic social teaching[37] and extending them to encompass a category that I am proffering as Catholic planetary teachings.[38]

Thinking about our species in relation to others and the systems of Earth within which we mutually function continues to consume a great deal of my time. That *Homo sapiens* emerged from other hominids approximately 200,000 years ago has been solidly established by evolutionary biologists and paleontologists. Also solidly established is the fact that our species has so adversely affected other species and systems of Earth that a new geological age has been identified and named ominously as the Anthropocene. It should humble and motivate all people to respond to Pope Francis's call for dialogue and action. Perhaps a model of ourselves that is sufficiently descriptive of the type of behavior we should be demonstrating will help. Toward that possibility, I was inspired by Aquinas's thinking to develop a model of the human as the virtuous cooperator who steadfastly cooperates prudently, justly, moderately, humbly, and compassionately with other diverse species and systems to constitute a good, beautiful, and cohesive creation that God empowers to function without interference, sustains in existence, and calls to completion.[39]

Conclusion

My encounter with Earth spans feelings of awe prompted by natural phenomena that God made possible, disgust when realizing the degradation and destruction that human activities are causing, a mix of frustration and satisfaction when striving with others to advocate constructive actions and policies, delight when discovering sources in the Catholic theological tradition

37. The life and dignity of the human person, the rights and responsibilities of persons, the call to family, community and participation, the dignity of human labor and rights of workers, the preferential option for poor people, the solidarity of all people, their common good, and the principle of subsidiarity for decision-making. Care for God's creation was the latest principle of Catholic social teaching added for the faithful to consider. Brief explanations of these principles are available at http://www.usccb.org/beliefs-and-teachings/what-we-believe/catholic-social-teaching/seven-themes-of-catholic-social-teaching.cfm.

38. After having briefly shared "Developing an Ecological Consciousness," at Lumen Christi Institute's Conference on Caring for our Common Home: Economics, Environment, and Catholic Social Thought held in Chicago on May 20, 2016, I am working on a book-length monograph.

39. Schaefer, "The Virtuous Cooperator."

for addressing ongoing problems, and thrill in having the opportunity at Marquette University to share with my students compelling religious and spiritual sources that empower and spur them to facilitate the flourishing of Earth. Apparent to me throughout has been the realization that scientific knowledge is essential for making decisions about how to act. Increasingly apparent and important has been the need for reflecting theologically about our species, Earth, and God informed by our current understanding of the world to achieve the deepest possible meaning for living responsibly. From my research has emerged a model of the human as the virtuous cooperator who acts steadfastly prudent, just, moderate, courageous, humble, and compassionate toward other people and species within systems of Earth for our mutual flourishing and desire to cooperate with God.

Bibliography

Agency for Toxic Substances and Disease Registry, US Department of Health. "Toxic Substances Portal—Polychlorinated Biphenyls (PCBs)." https://www.atsdr.cdc.gov/toxfaqs/tf.asp?id=140&tid=26.

Benedict XVI. *If You Want to Cultivate Peace, Protect Creation*. Message on the 2010 World Day of Peace. http://w2.vatican.va/content/benedict-xvi/en/messages/peace/documents/hf_ben-xvi_mes_20091208_xliii-world-daypeace.html.

Day, Jennifer. "International Joint Commission Releases Special Report on Pollution Cleanup Efforts in Great Lakes Areas of Concern." International Joint Commission, March 25, 1998. http://www.ijc.org/en_/news?news_id=353.

Department of Natural Resources, State of Wisconsin. "The Sheboygan River Remedial Action Plan." http://dnr.wi.gov/topic/greatlakes/documents/SheboyganRiverRAPStage11989.pdf; http://dnr.wi.gov/topic/greatlakes/documents/SheboyganRiverRAPUpdate2014.pdf.

Francis. *Laudato sí, On Care for our Common Home*. Rome: Libreria Editrice Vaticana, 2015.

John Paul II. *Peace with God the Creator, Peace with All of Creation*. Message on the 1990 World Day of Peace. http://w2.vatican.va/content/john-paul-ii/en/messages/peace/documents/hf_jp-ii_mes_19891208_xxiii-world-day-forpeace.html (accessed 18 February 2016).

Leopold, Aldo. *A Sand County Almanac, and Sketches Here and There*. New York: Oxford University Press, 1949.

Office of the Great Lakes, Wisconsin Department of Natural Resources. "2015 Remedial Action Plan Update for the Sheboygan River Area of Concern." http://dnr.wi.gov/topic/greatlakes/documents/SheboyganAOCRAP2015.pdf.

Office of the Press Secretary, The White House. "FACT SHEET: Obama Administration Announces Actions to Ensure that Nuclear Energy Remains a Vibrant Component of the United States' Clean Energy Strategy." https://www.whitehouse.gov/the-press-office/2015/11/06/fact-sheet-obama-administration-announces-actions-ensure-nuclear-energy.

Paul IV. *Octogesima Adveniens*. Rome: Libreria Editrice Vaticana, 1971.

Public Service Commission of Wisconsin. Advance Plan Docket 05-EP-1. Madison, Wisconsin, August 17, 1978.

Schaefer, Jame. "Converting to and Nurturing an Ecological Consciousness—Individually, Collectively, Actively." In *Laudato sí, Background, Reception, and Commentary*, edited by Daniel DiLeo, 136–53.Winona, MN: Anselm Academic.

———. "Creating a Code of Ethics for the Ecosystem." Invited discussion paper, Science Advisory Board, International Joint Commission, Windsor, ON, 1991.

———. "Developing an Ecological Consciousness: Advancing Catholic Social Thought to Catholic Planetary Thought." Paper presented at Lumen Christi Institute's Conference on Caring for our Common Home: Economics, Environment, and Catholic Social Thought, Chicago, IL, May 20, 2016.

———. "Environmental Degradation, Social Sin, and the Common Good." In *God, Creation, and Climate Change: A Catholic Response to the Environmental Crisis*, edited by Richard W. Miller, 69–94. Maryknoll, NY: Orbis, 2010.

———. "Ethical Implications of Applying Aquinas's Notions of the Unity and Diversity of Creation to Human Functioning in Ecosystems." PhD diss., Marquette University, 1994.

———. "Imprudence and Intergenerational Injustice: Confronting the Ongoing Vices of Opting for Nuclear Fueled Electricity." *Environmental Ethics* 38.3 (Fall 2016) 259–86.

———. "State Opposition to Federal Nuclear Waste Repository Siting: A Case Study of Wisconsin 1976–1988." Center for Public Affairs, University of Wisconsin-Green Bay, Green Bay, WI, 1988.

———. *Theological Foundations for Environmental Ethics: Reconstructing Patristic and Medieval Concepts*. Washington, DC: Georgetown University Press, 2009.

———. "Toward a Code of Ethics for the Great Lakes Basin Ecosystem." Invited discussion paper, Science Advisory Board, International Joint Commission, Windsor, ON, 1989.

———. "The Virtuous Cooperator: Modeling the Human in an Age of Ecological Degradation." *Worldviews: Environment, Culture, Religion* 7.1–2 (2003) 171–95.

Schaefer, Jame, ed. *Confronting the Climate Crisis: Catholic Theological Perspectives*. Milwaukee, WI: Marquette University Press, 2011.

Schaefer, Jame, Michael E. Kraft, and Bruce B. Clary. "Politics, Planning, and Technological Risk: State and Citizen Participation in Nuclear Waste Management." Paper presented at the annual meeting of the Political Science Association, Chicago, IL, September 5, 1987.

U.S. Environmental Protection Agency. "Site Information for SHEBOYGAN HARBOR & RIVER." Accessed April 13, 2018. https://cumulis.epa.gov/supercpad/cursites/csitinfo.cfm?id=0505188.

———. "Great Lakes Restoration Initiative (GLRI)." Accessed April 13, 2018. https://www.epa.gov/great-lakes-funding/great-lakes-restoration-initiative-glri.

US Nuclear Regulatory Commission. "Rulemaking on the Storage and Disposal of Nuclear Waste." PR-50–51 (44 FR 61372). http://pbadupws.nrc.gov/docs/ML0336/ML033640136.pdf.

The Lord God Bird

Avian Divinity, Neo-Animism, and the Renewal of Christianity at the End of the World

—MARK I. WALLACE—

God on the Wing

This spring in the Crum Woods—on the edge of the Swarthmore College campus where I live and work—I encountered North America's largest wood-boring bird, the pileated woodpecker. A strikingly beautiful winged creature, this woodpecker is a large crow-sized bird, and brilliantly colored—mostly jet-black with slashing white stripes and a flaming red crest on top of its head. I first saw a pileated hammering out a nest hole in a dead tree on the banks of the Crum Creek. It cut wood chips out of the tree and then flicked them away from the newly excavated cavity with relished abandon, many of them splashing into the creek below. I watched this bird and its companion often feeding below the nest—pileated woodpeckers like ants and termites—by banging repeatedly into nearby trees. This is a loud, brash, in-your-face bird that lights up the forest with its soaring flight, screeching call, and head-bashing impact. After a couple of weeks of nest building by the adults, wonderfully, two baby pileated woodpeckers appeared. They called out and clattered around the nest just like the parents. Another two weeks passed, and the babies fledged—effortlessly, it seemed, as they took flight through the forest echoing their parents' wild energy and noisy shrieks.

Ornithologist Scott Weidensaul writes about the reactions local people often have had at catching sight of the pileated woodpecker or its almost identical cousin, the ivorybill woodpecker, which is now almost certainly extinct. Historically, this spectacle prompted observers to call these sensational creatures the "By-God, look at that bird" woodpeckers—or simply, in moments of gasping, exclamatory wonder, "the Lord God bird." Weidensaul writes

> The names [country folk] used for the ivorybill reflected the dazzle of seeing one of these huge birds rowing through the light-splashed swamps on powerful wings. King of the Woodpeckers, they called it. Log-cock. King Woodchuck. Giant woodpecker. Log God. Like the smaller but similar pileated woodpecker, it was sometimes called the Lord God bird, or the By-God, because that's what a breathless greenhorn said when he first saw one: By God, look at that bird.[1]

To be sure, such outbursts can be read simply as impulsive reactions to the sight of large, multicolored woodpeckers—the ivorybill or the pileated—ripping apart old logs with their massive bills, or flying through the forest, blackening out the sun's light overhead. But could such upwellings of emotion also be read as a hint of something else? Is it possible such outpourings are a clue in our core psyches that we are aware of something awe-inspiring—even divine—at play in the forests that surround us?

For me, when I saw the pileated woodpecker leave its nest and soar through the trees, darkening the ground at my feet with its almost thirty-inch wingspan, I also exclaimed, "Oh, my God!" and I too felt I had met face to face with a sort of avian deity. I wondered whether my response to this feathered God overhead was not unlike the reaction of Jesus' onlookers to the aerial Spirit above them at the time of Jesus' baptism in the Jordan Valley, two millennia ago. I asked myself, could the Holy Spirit, the winged God at Jesus' baptism, once more be taking flight as the Lord God bird in the Crum Woods?

In spite of my chance meeting of the Crum Woods bird God—in a manner analogous to Jesus' disciples' encounter with the winged Spirit at the Jordan River long ago—Christianity today is largely viewed as an unearthly religion with little to say about everyday life in the natural world. Because it has focused so heavily on the salvation of human souls, it has lost touch with the role the verdant world of animals and plants, land and water, plays in the spiritual well-being of humankind and otherkind. In principle,

1. See Weidensaul, *The Ghost with Trembling Wings*, 49, and his discussion of the historical depredation of, and contemporary search for, the likely extinct "ghost bird," the ivorybilled woodpecker, and the remarkable recovery, due to habitat restoration, of its nearly identical cousin, the pileated woodpecker.

Christian belief in the incarnation of God in the human Jesus renders biblical faith a fleshy, this-worldly belief system. In reality, however, Christianity is best known for its war against the flesh by denigrating bodily impulses as a source of temptation, and by dismissing the material world as contaminated by sin and inimical to humans' destiny in a far-removed heaven of bodiless bliss. This essay argues that this picture of Christianity as unearthly and hostile to the creaturely world, while historically accurate, misses the supreme value biblical religion assigns to all of the denizens of God's good creation, human and more-than-human alike.

Christian Animism

I maintain that the picture of a Christianity hostile to Earth misses the startling portrayals of God as the beaked and feathered Holy Spirit, the third member of the Christian Trinity who, alongside the Father and Son, is the "animal God" of historic Christian witness.[2] Appearing in the Christian Scriptures as a winged creature at the time of Jesus' baptism, the bird-God of the Bible signals the deep grounding of faith in the natural world. But due to the bias of world-denying Christianity, this reality of God in creaturely form—not only in the form of the human Jesus, but also in the form of the birdy Spirit—has been missed by most Christian thinkers and practitioners. I wish to correct this oversight and pave the way for a new Earth-loving spirituality grounded in the ancient image of God as an avian life-form.

Throughout the history of Christianity and church art, the Holy Spirit has soared through the sky—in bright plumage and airy flesh—in altarpieces, pulpit ornaments, illuminated manuscripts, rood screens, and in religious painting and statuary in general. Trinitarian portrayals of the Spirit eloquently make this point: the Father and Son are depicted in human familial terms, respectively, while the Spirit is figured as the winged divinity who mediates the relationship of the other two members of the Godhead. My recovery of God's animal body within biblical and Christian sources might at first be startling, even sacrilegious, for some readers. Even though the Bible speaks directly about God as Spirit becoming a feathered creature ("When Jesus was baptized, the Holy Spirit descended upon him in bodily form as a dove" [Luke 3:21–22]), religion and biblical scholars alike have oftentimes dismissed the descriptions of God's Spirit as a bird in the New

2. Recent work on the Spirit that has informed my own thinking includes Betcher, *Spirit and the Obligation of Social Flesh*; Wallace et al., "Spirit"; Derrida, *Of Spirit*; and Holl, *The Left Hand of God*.

Testament as a figure of speech, and do not regard this and similar texts as actual descriptions of the avifauna God became and is becoming. Nevertheless, I contend here that the full realization of Christianity's historic self-definition as a scriptural, incarnational, and trinitarian belief system is *animotheism*[3]—the belief that all beings, including nonhuman animals, are imbued with God's presence. Buried deep within the subterranean strata of the Christian witness is a trove of vibrant bodily images for God in animal form (as well as in human and plant forms), including, and especially, the image of the avian body of the Holy Spirit. Woven into the core grammar of Christian faith, then, is the belief in the Spirit as the *animal* face of God, even as Jesus is the *human* face of God.

I call this new but ancient vision of the world "Christian animism" in order to signal the continuity of biblical religion with the beliefs of indigenous communities that God or Spirit enfleshes Godself within everything that grows, walks, flies, and swims, on, in, and over the great gift of creation.[4] But labeling Christianity as an animist belief system—the conviction that all things, including so-called inanimate objects, are alive with sacred presence and worthy of human beings' love and protection—is odd for Christian believers and religious scholars alike, who regard biblical religion as diametrically opposed to the pagan religions of primordial people. Historically, traditional Christianity viewed itself as a divinely inspired religion of the book that is categorically different from religions that showed special regard for sacred animals, tree spirits, revered landscapes, and hallowed seasons of the year. In this telling, Christianity replaced the old gods of pre-Christian animism with the new revealed religion of Jesus, the saints, and the Bible. In its self-definition, Christianity regarded itself as an otherworldly faith that superseded heathen superstitionism insofar as its focus was on an exalted and unseen Deity who is not captive to the vicissitudes of mortal life on Earth.

Today, innovative attempts at forging connections between biblical religion and primordial belief systems marks a sea change away from earlier comparativist studies of "revealed religions" such as Christianity vis-à-vis preliterate religious cultures. Beginning in the late twentieth and early twenty-first centuries, a profound shift has taken place toward a critical understanding of the centrality of animal bodies and subjectivities in the formation and contemporary expression of *all* of the world's religions,

3. The term belongs to Mircea Eliade in *Shamanism*, 156–58.

4. In this regard, see White, "The Historic Roots of Our Ecologic Crisis"; Quinn, "Animism"; McGaa, *Mother Earth Spirituality*; Beck, *Christian Animism*; Deloria, "Sacred Places and Moral Responsibility"; and Grim, "Indigenous Traditions and Deep Ecology."

including Christianity.[5] In earlier studies, Christianity was championed as an evolved form of book-based monotheism over and against rival forms of so-called primitive or savage religions that were based in fertility rituals and animal sacrifice. The hoary opposition between pure monotheism and nature, and animal-based religion is bedrock to all of nineteenth- and early twentieth-century British anthropology of religion, including E. B. Tylor's *Primitive Culture*, William Robertson Smith's *The Religion of the Semites*, and James Frazer's *The Golden Bough* and *The Worship of Nature*. At the heart of this division between the primitive and the modern in early Victorian studies of religion, the notion of animism was deployed as a proxy for the benighted epistemologies of first peoples who envisioned the cosmos as an intersubjective communion of living beings, including animal beings, with shared intelligence, personhood, and communication skills.[6]

Sharing resonances with the Latin word *animus*, which means "soul" or "spirit," the idea of animism was significantly advanced in the modern West by the anthropologist E. B. Tylor, who used it to analyze how primordial people attributed "life," "soul," or "spirit" to things, living and nonliving. In *Primitive Culture*, Tylor says, quoting another theorist, that in animism "every land, mountain, rock, river, brook, spring, tree, or whatsoever it may be, has a spirit for an inhabitant; the spirits of the trees and stones, of the lakes and brooks, hear with pleasure . . . man's pious prayers and accepts his offerings."[7] Tylor's study of animism emerged out of an evolutionary, occidental mind-set that described, at least for Victorian readers, the unusual pan-spiritist beliefs and practices of first peoples—the ancient belief that all things are bearers of spirit. Tylor denigrated animism as the superstitious worldview of savage tribes whose beliefs eventually gave way, in his thinking, to the march of reason and science in civilized societies.[8] While the term is tainted by Tylor's colonial elitism (animism is characteristic of "low humanity" rather than "high culture"), the concept of animism today car-

5. For two comparative religious and theological works that analyze Christianity, among other religions, and sacred animals, in particular, see Waldau and Patton, eds., *A Communion of Subjects*, and Moore, *Divinanimality*.

6. As indigenous religions scholar John Grim writes, "During the late nineteenth century colonial period interpretive studies described communication with animals among indigenous peoples as a failed epistemology. The assumption that only humans know, or a least that only humans report on their knowing, resulted in the long-standing critique of indigenous ways of knowing coded in the term animism. As a means of actually knowing the world, animism was dismissed as simply a delusion, or a projection of a deluded human subjectivity." Grim, "Knowing and Being Known by Animals," 379.

7. Tylor, *Primitive Culture*, 2:170–71.

8. See, e.g., ibid., 1:385, where Tylor insists that "Animism characterizes tribes very low in the scale of humanity."

ries a certain analytical clarity by illuminating how indigenous communities, then and now, envision nonhuman nature as "ensouled" or "inspirited" with living, sacred power. As comparative religions scholar Graham Harvey writes, animism "might be summed up by the phrase "all that exists lives" and, sometimes, the additional understanding that "all that lives is holy."[9]

In Harvey's formulation of animism, nature is never dull and inert but inherently alive with the infusion of Spirit or spirits into all things. Here there is no distinction between living and nonliving, between animate and inanimate. Harvey's use of the phrase "all that exists lives" means that nature is not brute matter, but always full of life and animated by its movement, weight, color, voice, light, texture—and spiritual presence. Nature's capacity for *relatedness,* its proclivity to encounter the human subject, as the subject encounters it, in constantly new and ever-changing patterns of self-maintenance and skillful organization is the ground tone of its vibrant and buoyant power. As philosopher David Abram similarly argues, nature or matter is not a dead and lesser thing that stands in a lower relationship to animate spirit, but a self-organizing field of living, dynamic relationships.

> Yet as soon as we question the assumed distinction between spirit and matter, then this neatly ordered hierarchy begins to tremble and disintegrate. If we allow that matter is *not* inert, but is rather animate (or self-organizing) from the get-go, then the hierarchy collapses, and we are left with a diversely differentiated field of animate beings, each of which has its own gifts relative to the others. And we find ourselves not above, but in the very midst of this living field, our own sentience part and parcel of the sensuous landscape.[10]

Abram and others analyze how indigenous peoples celebrated, and continue to celebrate, relations with other-than-human communities of beings that are alive with spirit, emotion, desire, and personhood. This ascription of personhood to all things locates human beings in a wider fraternity of relationships that includes "bear persons" and "rock persons" along with "human persons."[11] At first glance, this is an odd way to think, since Western ontologies generally divide the world between human persons, other animals, and plants as *living things*, on the one hand, and other things such as earthen landscapes, bodies of water, and the airy atmosphere as *non-sentient*

9. Harvey, *Animism*, 1:81.

10. Abram, *Becoming Animal*, 47.

11. See "new animism" studies of human-nature intersubjectivity in Curry, "Grizzly Man and the Spiritual Life"; Stuckey, "Being Known by a Birch Tree"; and Abram, *Becoming Animal.*

elements, on the other. Animism flattens these distinctions along a continuum of multiple intelligences: now everything that *is* is alive with personhood and relationality, even sentience, according to its own capacities for being in relationship with others. As Harvey says, "Animists are people who recognize that the world is full of persons, only some of whom are human, and that life is lived in relationship with others."[12] *All* things are persons, only *some* of whom are human, because *all* beings are part of an intersubjective community of relationships, only some of whom are recognizable as *living* beings by us.

In general, however, most scholars of religion regard animism as far removed from Christianity, both culturally and theologically. Comparative religions scholar Bron Taylor, for example, writes that in spite of attempts to bring together animism, which he calls "dark green religion," and the major world religions, such as Christianity, these traditions have different origins, share different worldviews, and cannot genuinely cross-pollinate. He writes, "For the most part, in spite of occasional efforts to hybridize religious traditions, most of the world's major religions have worldviews that are antithetical to and compete with the worldviews and ethics found in dark green religion."[13] I will argue the contrary, namely, that while the Christian religion largely evolved into a sky-God tradition forgetful of its animist origins, its carnal identity is paradigmatically set forth in canonical stories about the human body of the historical Jesus, on the one hand, and, provocatively, the animal body of the avian Spirit, on the other. Writing as a theologian, my reading of Christian history and the biblical texts cuts against the received misunderstanding of Christianity as a discarnational religion, so to speak, and attempts to return biblical faith to its true animist beginnings and future prospects.

The Bird God

Many people of faith, and especially Christians, are used to speaking of the *humanity* of God, but are less comfortable speaking about the *animality* of God. As some scholars suggest, perhaps this is the case because the animal

12. Harvey, *Animism*, xi.

13. Taylor, *Dark Green Religion*, 178. In a footnote to this quote, Taylor notes my earlier work in animist Christianity in *Finding God in the Singing River*, as "an exception that proves the rule" that Christianity and animism are categorically distinct. But if there are exceptions to a rule, the rule itself should be questioned. In my judgment, the age-old opposition between Christianity and animism trades on a false choice.

world appears relatively insignificant in the Christian Scriptures. As Laura Hobgood-Oster puts it in her otherwise luminous analysis of animals and Christianity in *Holy Dogs and Asses: Animals in the Christian Tradition*, "Although animals are not prominent in either the canonical or the extra-canonical gospels, powerful stories emerge from the relatively unknown extracanonical traditions."[14] My point is the opposite, namely, that the nonhuman animal plays a central and commanding role in the canonical drama of the New Testament *just insofar as* God in Godself is consistently depicted as a beaked and feathered being in the person of the Holy Spirit. Focusing on the biblical beginnings of Jesus' adult life elucidates the importance of the airborne Holy Spirit in the inauguration of Jesus' public work. With special reference to Jesus' baptism, a return to the textual sources of Christian origins sets forth the critical part played by God's winged Spirit in the formation of Jesus' ministry.

In the story of Jesus' baptism in the canonical Gospels—and in one unauthorized Gospel called *The Gospel of the Ebionites*[15]— the Spirit comes down from heaven as a bird and then alights upon Jesus' newly baptized body. All five accounts narrate the same gospel memory, namely, that as Jesus presents himself to be baptized by John the Baptist, and is baptized, the Spirit descends upon Jesus as a dove from heaven as a voice from heaven says, "This is my beloved son with whom I am well pleased." At the time of Jesus' baptism, it seems certain that the power and wonder of the descending Spirit-bird, along with a heavenly voice, indelibly seared the memories of the authors of each of the five Gospels. This collective memory of the feathered divine being appearing at the debut of Jesus' public ministry must have etched a lasting image in the minds of each of the canonical and extra-canonical authors, all of whom told, roughly speaking, the same story.

On one level, I suspect that the people who came to John for baptism were not surprised to see the Holy Spirit in the form of a dove. In biblical times, doves—in addition to other divinized flora and fauna—figured prominently in the history of Israel as archetypes of God's compassion.[16] On another level, however, Jesus' audience at the time of his baptism must have wondered, "Is God the Spirit really a feathered creature?" In all five of our

14. Hobgood-Oster, *Holy Dogs and Asses*, 47.

15. See "The Gospel of the Ebionites," in Ehrman, *Lost Scriptures*.

16. Noah sends a dove out after the flood (Gen 8:6–12). Abraham sacrifices a dove to God to honor the covenant (Gen 15). Solomon calls his beloved "my dove," a heartfelt term of longing and endearment (Song of Solomon 2:14, 4:1, 5:2, 6:9). Ezekiel and Jeremiah refer to the doves' swift flight, careful nesting, and plaintive cooing (Ezek 7:16; Jer 48:28). As divine emissary and guardian of sacred order, the dove is a living embodiment of God's protection, healing, and love.

texts, the Greek term used to denote this winged being is *peristera*. Interestingly, in Greek, as in English, the name for this sort of bird can mean either "dove" or "pigeon," a confusion that is papered over by the New Testament translations, as well as by later artistic depictions of this bird as a beautiful and gently ethereal white dove. But the original Greek word for this creature—*peristera*—has a wider semantic range and can mean either one of these two types of birds, dove or pigeon.[17] In English taxonomy as well, both the dove and the pigeon constitute the same "branch" on the avian "tree" that ornithologists use to classify birds of all types.

In today's common terminology, this one bird with different names is alternately referred to as the *rock dove*, the *rock pigeon*, or the *domestic pigeon*. All of these names refer to the same exact bird. Historically, scholars think the rock dove was the aboriginal name for this bird, the iconic heavenly dove referred to in the Gospel texts in question. Today, the rock dove, rock pigeon, or domestic pigeon can be found on every continent (except Antarctica), many oceanic islands, and almost every habitat strewn across the Earth. It is as likely, then, that God's avian Spirit in the Gospels was as much a rambunctious feral pigeon as it was a virginal all-white dove.[18]

Of all the Gospel accounts, Luke's story of Jesus' baptism is a particularly resonant summary of the Gospels' overall narrative of Jesus' ritual immersion. Luke writes: "Now when all the people were baptized, and when Jesus also had been baptized and was praying, the heaven was opened, and the Holy Spirit descended upon him in bodily form, as a dove, and a voice came from heaven, 'Thou art my beloved Son; with thee I am well pleased'" (Luke 3:21–22). After highlighting Jesus' baptism by John, and the opening of the heavens, Luke says, "the Holy Spirit descended upon [Jesus] in bodily form (*somatiko eidei*), as a dove/pigeon (*hos peristeran*)" (Luke 3:22). Here the Greek phrase *somatiko eidei* means "in bodily form" or "in bodily essence." In this phrase, the Greek adjective *somatikos*, from the noun *soma* (body), signifies the shape or appearance of something in corporeal form. The point here is that God as Spirit is fully carnal and creaturely. Moreover, God as Spirit is fully creaturely in the form of an animal body, the dove/pigeon of the Gospel witness. The Holy Spirit, the third member of the Godhead, enters existence in *animal* form—even as the second member of the Godhead, Jesus, enters existence in *human* form.

The particular beaked-and-feathered body Luke's spirit-animal has become is defined by the phrase *hos peristeran*, which means "as/even like/

17. To this point, see W. Stewart McCullough's entries on *peristera*, which he alternately translates as "dove" and "pigeon," in his articles by these same names, in Buttrick, ed., *The Interpreter's Dictionary of the Bible*, 1:866–67, 3:810.

18. For more on this position, see Blue, *Consider the Birds*, 9–10.

just as a dove/pigeon"—that is, the Spirit's body is thoroughly birdlike. Some English translations of the Lukan and other Gospel accounts of Jesus' baptism miss this point. While the Revised Standard Version says, "The Holy Spirit descended upon him *as* a dove," the New Revised Standard Version prefers, "The Holy Spirit descended upon him *like* a dove" (emphases mine).

But the preposition *hos*—from *hos peristeran* in the original Greek text of Luke 3:22 and elsewhere—does not operate here metaphorically or analogically, but appositionally. The expression "as a dove/pigeon (*hos peristeran*)" in this context is not a simile that says that the Spirit descended in bodily form *like* a dove/pigeon, but is rather an explanatory phrase that describes the *actual* physical being the Spirit has become. In other words, the Spirit descended in bodily form *as* a dove/pigeon—or in the technical language of Mircea Eliade, the Spirit is *ornithomorphic*.[19] In the grammar of predication, the Spirit *is* a dove/pigeon, not *like* a dove/pigeon. Luke 3:22 is not, then, a figure of speech to connote the passing birdlike appearance of the Spirit in this one instance, but a literal description of the actual bird-God the Spirit has become. God *is* a bird in the Gospel of Luke.

In my judgment, then, the Gospels' portrait of this winged divine animal champions the eternal unity of all things: God and Earth, spirit and flesh, divine life and birdlife, divinity and animality. The point of the baptism narrative cycle is that the Gospels' heavensent dovey pigeon is God enfleshing Godself in carnal form, but now not only in human flesh in the person of Jesus (God's Son), but also in animal flesh in the person of the Spirit (God's Spirit). Could it be then, glossing the Nicene Creed, that the Holy Spirit, the one who with the Father and Son is worshipped and glorified, is a lowly, ordinary pigeon—so much so that when we look into the black pupils and brightly colored irises of a common pigeon, we are looking into the face of God, the Lord and giver of life?

Plumbing the Depths of Subscendence

This vision of an avian God challenges traditional understandings of Christian origins as fundamentally opposed to animism. To maintain this opposition, the standard paradigm for understanding Christianity focuses on a faraway time when Christian civilization supposedly vanquished ancient paganism. In antiquity, so goes the argument, every living thing was experienced as alive with its own guardian spirit until Christian missionaries arrived to destroy the indigenous cultures and beliefs of primordial people.

19. See Eliade, *Shamanism*, 156–58.

Historian Lynn White Jr., arguably the fountainhead of today's emerging academic field of religion and ecology, advanced this myth of Christian beginnings in his seminal article, "The Historical Roots of Our Ecologic Crisis." White writes,

> Christianity, in absolute contrast to ancient paganism and Asia's religions (except, perhaps, Zorastrianism), not only established a dualism of man and nature but also insisted that it is God's will that man exploit nature for his proper ends. At the level of the common people this worked out in an interesting way. In Antiquity every tree, every spring, every stream, every hill had its own genius loci, its guardian spirit. These spirits were accessible to men, but were very unlike men; centaurs, fauns, and mermaids show their ambivalence. Before one cut a tree, mined a mountain, or dammed a brook, it was important to placate the spirit in charge of that particular situation, and to keep it placated. By destroying pagan animism, Christianity made it possible to exploit nature in a mood of indifference to the feelings of natural objects.[20]

White's formulation of the Christian-animist opposition is generally accepted by most scholars of religion today, and by Christian practitioners themselves, but this opposition rests on a simple fallacy: an acceptance of the guardians of Christian orthodoxy's self-understanding of biblical faith's triumph over its pagan, animist origins. Contemporary religious studies scholars, as well as Christianity's most ardent defenders, operate from this anti-Christian animist mind-set. In the case of the defenders, evangelical apologist Pat Zukeran makes this opposition clear when he writes, "From Genesis to the present the biblical world-view has clashed with the world-view of animism."[21] Similarly, conservative theologian Robert A. Sirico opines that "the whole of the religious tradition of the West," in its victory over animism, now argues "that nature has no soul and is not in the salvific plan of God."[22]

Similar to these defenders of Christian traditionalism, and akin to religion scholars such as Taylor, White's setup of the triumph of Christianity over animism is inattentive to historical complexity and contemporary nuance. To be sure, Western history is replete with images of prominent holy men who, for example, cut down groves of sacred trees, among other destructive acts, in order to desacralize nature among pre-Christian peoples.

20. See White, "The Historical Roots of our Ecologic Crisis," 1205.

21. Zukeran, *World Religions Through a Christian Worldview*, 43.

22. Sirico, "The New Spirituality."

But while the axe-wielding saint is a powerful and disturbing anti-environmental image in Christian history, this trope only tells one side of the story. As theologian Belden Land writes, we are all familiar with legends telling

> once again the defeat of the natural world through the power of the cross. This all-too-frequently repeated narrative distorts the actual importance of trees, and of nature generally, in the history of Christian spirituality. For every story about saints who cut down trees in an act of anti-pagan triumphalism, there are two stories of saints living in hollow oaks, singing the holy office along with their arboreal friends, even causing the trees to burst into leaf in deep midwinter. If St. Martin of Tours allowed himself to be bound to a stake in the path of a falling sacred pine (though on being cut, of course, it fell in the opposite direction), Saints Gerlach, Bavo, and Vulmar were celebrated for living in hollowed-out trees, St. Victorinus for causing a dead tree to blossom at his death, and St. Hermeland for driving caterpillars from the forest she loved.[23]

Images of the crusading saint with axe in hand makes for good newspaper copy, but it says little about the thoughtful efforts by Christians and other people of faith who valorized in the past, and do so today, the Earth as a blessed community of living beings deserving of our respect and reverence.[24] Thus, in the contest between emerging Christian and indigenous animist societies, White et al. fail to comprehend the subtly organic and evolutionary nature of the clash between, and at times the marriage of, these two cultural worldviews—an evolution where differences were negotiated instead of being frozen into polar opposition.

Far from Christianity exercising victory over animism in its efforts to exploit the living world, the actual beginnings of Christian faith are rooted in the deep soil of an animist vision of interconnected, sacred nature. At its core, *pace* White and many others, Christianity did not *annihilate* animism in its putative dualistic theology, but rather *sublated* it to its articulation of the one God now enfleshed within this world through the human person of Jesus (and, as I've suggested, through the animal person of the Spirit). By saying Christianity sublated animism I do not mean that it subordinated

23. Lane, *Ravished by Beauty*, 125.

24. To note two examples in this regard, see Roskos, "Felling Sacred Groves," on Christian groups dedicated to land and forest conservation, for example, Christians for the Mountains, The Religious Campaign for Forest Conservation, and the Natural Religious Partnership for the Environment; and regarding historic attempts by Quakers to save tens of thousands of acres of first- and second-growth rain forest in Costa Rica, see Nadkarni and Wheelwright, *Monteverde*.

animism to itself, but, rather, that it preserved animism's reverence for sacred nature within the horizon of its own incarnational belief system. This sensibility is present, for example, in Paul's sermon in Athens when, quoting the spiritual writers of his day, he preaches that the "God who made the world and everything in it . . . indeed, he is not far from each one of us. For 'In him we live and move and have our being'; as even some of your own poets have said, 'For we too are his offspring'" (Acts 17:24, 27–28). Here Paul critically correlates the Greek philosophical idea that all things are *in* God with the Jewish and Christian story that the one God of heaven and earth *created* all things. The God who is one is also the God within whom the many subsist. In Paul's theology, the God of all creation, according to the biblical witness, is now as well the all-encompassing divinity within which we live and move and have our being, in the manner of pan*en*theistic animist spirituality.

Formative Christianity, therefore, both preserved and transposed animist sensibilities into a new and emerging biblical idiom. This transposition was played against the incarnational baseline of Christian discourse vis-à-vis its contestation and cohabitation with animist cultures. In this regard, nascent Christianity both transformed, and was transformed by, primordial Earth-based religion. My reference here to Paul's panentheism in Acts 17, and my overall focus on the Holy Spirit as an avian form of divinity, make this point in the biblical register. But one can just as easily make this point with reference to other New Testament animist images that were popular in the immediate aftermath of the post-biblical period in Christian history.[25]

25. In this vein, one could explore "The Martyrdom of Polycarp," a mid-second century CE avian spirit possession story focusing on the Christian martyr Polycarp. This is an important animist "bridge" text in Christian history that links theologically the close of the New Testament and the emergence of the writings of the apostolic fathers. In the account, Polycarp is commanded to worship the gods of the Roman state under orders of the area proconsul. When he refuses, he is lashed to a stake and prepared for a funeral pyre. At this juncture, a great billowing ball of fire is set around Polycarp by his enemies. When the fire fails to kill Polycarp, an executioner stabs him with a dagger. "When he did so, *a dove came forth,* along with such a quantity of blood that it extinguished the fire, striking the entire crowd with amazement that there could be so much difference between the unbelievers and the elect." Ehrman, "The Martyrdom of Polycarp," 187 (my emphasis). Texts such as "The Martyrdom of Polycarp" are not predicated on a war between biblical and natural religion but, rather, on a mutually transformative dialectic between the Christian tropes of faith and martyrdom, on the one hand, and earth religion motifs such as fire, flesh, blood, and avian divinity, on the other.

Theologically Crossing the Species Divide

Many of us today are enduring the shock and profound loss of living in the period of what some are calling "the anthropocene," "the end of nature," or the time of the "sixth great extinction."[26] Due to anthropogenic climate change, the contemporary loss of plants and animals appears to be 10,000 times greater than naturally occurring extinction rates, with tens of thousands of species lost annually. The hopelessness that characterizes our looming dread about fracturing biodiversity—Greenland lost a trillion tons of sea ice in just the past four years alone due to global warming[27]—leaves many of us unsteady and fearful about the future. At times, historic ascetic Christianity has contributed to this feeling of doom by making war against the natural world and the flesh, and by denying to God a place in the world, making the living world a dead place, an empty place, a godless place.[28]

But this emphasis runs aground on the shoals of the biblical animist belief in the living goodness of all inhabitants of sacred Earth. At the end of the world, the hope for a renewal of Christian faith rests on its practitioners regaining a profound sense of God's fleshy, this-worldly identity. Philosopher Gayatri Spivak writes that "animist liberation theologies" could be the next political innovation on the horizon following Christianity's recent turn to liberation and ecology.[29] If Spivak is right, then animism is ready to be recovered as the lost treasure of Christian belief—and the ground tone of its revivial today. In spite of Christianity's erstwhile anthropocentric chauvanism, God is not, in a neo-animist worldview, a bodiless heavenly being divorced from the material world, but manifested instead as the winged Spirit who protectively alights on Jesus at the time of his baptism, and thereby infuses the world with sustaining love. Ironically, in the light of its misunderstood history, Christianity is a religion of *carnal subscendence*, of God giving Godself in the many life-webs of existence, not *other-worldly transcendence*, where God is pictured as cosmically remote from planetary well-being.

This, then, is the great hope of Christian animism: that through the eyes of faith, the world will be experienced again as teeming with living persons—tree people, rock people, river people, bird people, including, for

26. Kolbert, *The Sixth Extinction*.

27. Harvey, "Greenland Lost a Staggering 1 Trillion Tons of Ice in Just Four Years."

28. See this extracanonical exhortation to Christian asceticism from late antiquity, urging its readers to cleanse themselves of worldly pollution by overcoming fleshly temptations: "Blessed are those who have not polluted their flesh by craving for this world, but are dead to the world that they may live for God!" Ehrman, "Pseudo-Titus."

29. Spivak, *A Critique of Postcolonial Reason*, 355 n. 59.

me, the pileated woodpecker—all of whom urge us to comport ourselves to them with dignity and compassion. In *Paradise Lost*, John Milton writes that "millions of spiritual creatures walk the earth unseen, both when we wake and when we sleep."[30] Is Milton right? Could this be so? Could the world for us again be envisioned and enlivened as a living place full of spiritual creatures, and thereby deserving of our reverence, and worthy of our protection? If this could be so, then into the living waters, vital food chains, fertile soil, and the quickening atmosphere everywhere, God would be seen as streaming forth to insure life and vitality for all beings. If this could be so, then we would care for the world with abandon—even as God too would be understood *kenotically* as holding nothing back—the bird-God who pours out Godself into the wonder and plenitude of our planetary commons. If this could be so, then all things would be viewed as bearers of the sacred; each and every creature would be seen as a portrait of God; and everything that is would be cherished as holy and good.[31]

Bibliography

Abram, David. *Becoming Animal: An Earthly Cosmology*. New York: Vintage, 2010.

Beck, Shawn Sanford. *Christian Animism*. Alresford, Hants, UK: Christian Alternative, 2015.

Betcher, Sharon V. *Spirit and the Obligation of Social Flesh: A Secular Theology for the Global City*. New York: Fordham University Press, 2013.

Blue, Debbie. *Consider the Birds: A Provocative Guide to Birds of the Bible*. Nashville: Abingdon, 2013.

Buttrick, George Arthur, ed. *The Interpreter's Dictionary of the Bible*, 5 vols. Nashville: Abingdon, 1962.

Curry, Patrick. "Grizzly Man and the Spiritual Life." *Journal for the Study of Religion, Nature and Culture* 4 (2010) 206–19.

Deloria, Vine Jr. "Sacred Places and Moral Responsibility." In *Worldviews, Religion, and the Environment: A Global Anthology*, edited by Richard C. Foltz, 83–91. New York: Wadsworth, 2002.

Derrida, Jacques. *Of Spirit: Heidegger and the Question*. Translated by Geoffrey Bennington and Rachel Bowlby. Chicago: University of Chicago Press, 1989.

Eliade, Mircea. *Shamanism: Archaic Techniques of Ecstasy*. Translated by Willard R. Trask. New York: Pantheon, 1964.

30. John Milton, *Paradise Lost* bk. 4, v. 678.

31. Here I am reminded of the animist sensibility in the Book of Job, where God says, "But ask the animals and they will teach you; the birds of the air and they will tell you; speak to the Earth and it will teach you; and the fish of the sea will declare to you" (Job 12:7–8).

Ehrman, Bart D. "The Gospel of the Ebionites." In *Lost Scriptures: Books that Did Not Make It into the New Testament*, 12–14. Oxford: Oxford University Press, 2003.

———. "The Martyrdom of Polycarp." In *The New Testament and Other Early Christian Writings*, 183–88. Oxford: Oxford University Press, 2004.

———. "Pseudo-Titus." In *Lost Christianities: The Battle for Scripture and the Faiths We Never Knew*, 239. Oxford: Oxford University Press, 2003.

Grim, John A. "Indigenous Traditions and Deep Ecology." In *Deep Ecology and World Religions: New Essays on Sacred Ground*, edited by David Landis Barnhill and Roger S. Gottlieb, 35–57. New York: SUNY, 2001.

———. "Knowing and Being Known by Animals." In *A Communion of Subjects: Animals in Religion, Science, and Ethics*, edited by Paul Waldau and Kimberley Patton, 373–90. New York: Columbia University Press, 2006.

Harvey, Chelsey. "Greenland Lost a Staggering 1 Trillion Tons of Ice in Just Four Years." *Washington Post*, July 19, 2016.

Harvey, Graham. "Animism—A Contemporary Perspective." In *The Encyclopedia of Religion and Nature*, edited by Bron R. Taylor et al., 2 vols., 78–83. New York: Continuum, 2005.

———. *Animism: Respecting the Living World*. New York: Columbia University Press, 1976.

Hobgood-Oster, Laura. *Holy Dogs and Asses: Animals in the Christian Tradition*. Urbana, IL: University of Illinois Press, 2008.

Holl, Adolf. *The Left Hand of God: A Biography of the Holy Spirit*. Translated by John Cullen. New York: Doubleday, 1998.

Kolbert, Elizabeth. *The Sixth Extinction: An Unnatural History*. New York: Henry Holt and Company, 2014.

Lane, Belden C. *Ravished by Beauty: The Surprising Legacy of Reformed Spirituality*. Oxford: Oxford University Press, 2011.

McGaa, Ed Eagle Man. *Mother Earth Spirituality: Native American Paths to Healing Ourselves and Our World*. New York: Harper and Row, 1990.

Moore, Stephen D., ed. *Divinanimality: Animal Theory, Creaturely Theology*. New York: Fordham University Press, 2014.

Nadkarni, Nalini M., and Nathaniel T. Wheelwright. *Monteverde: Ecology and Conservation of a Tropical Cloud Forest*. Oxford: Oxford University Press, 2000.

Quinn, Daniel. "Animism—Humanity's Original Religious Worldview." In *The Encyclopedia of Religion and Nature*, edited by Bron R. Taylor et al., 2 vols., 1:81–83. New York: Continuum, 2005.

Roskos, Nicole A. "Felling Sacred Groves: Appropriation of a Christian Tradition for Antienvironmentalism." In *Ecospirit: Religions and Philosophies for the Earth*, edited by Laurel Kearns and Catherine Keller, 483–92. New York: Fordham University Press, 2007.

Sirico, Robert A. "The New Spirituality." *The New York Times Magazine*, November 23, 1997, https://actonorg/node/6426.

Spivak, Gayatri Chakravorty. *A Critique of Postcolonial Reason*. Cambridge, MA: Harvard University Press, 1999.

Stuckey, Priscilla. "Being Known by a Birch Tree: Animist Refigurations of Western Epistemology." *Journal for the Study of Religion, Nature and Culture* 4 (2010) 182–205.

Taylor, Bron. *Dark Green Religion: Nature Spirituality and the Planetary Future.* Berkeley, CA: University of California Press, 2010.

Tylor, E. B. *Primitive Culture,* 2 vols. New York: Gordon, 1871, 1974.

Waldau, Paul, and Kimberley Patton, eds. *A Communion of Subjects: Animals in Religion, Science, and Ethics*. New York: Columbia University Press, 2006.

Wallace, Mark I. *Finding God in the Singing River: Christianity, Spirit, Nature.* Minneapolis: Fortress, 2005.

Wallace, Mark I., et al. "Spirit." In *Constructive Theology: A Contemporary Approach to Classical Themes*, edited by Serene Jones and Paul Lakeland, 239–78. Minneapolis: Fortress, 2005.

Weidensaul, Scott. *The Ghost with Trembling Wings: Science, Wishful Thinking, and the Search for Lost Species*. New York: North Point, 2002.

White, Lynn, Jr. "The Historic Roots of Our Ecologic Crisis." *Science* 155 (1967) 1203–7.

Zukeran, Pat. *World Religions Through a Christian Worldview*. Richardson, TX: Probe Ministries, 2008.

The Human Quest to Live in a Cosmos

—HEATHER EATON—

"The cosmos is within us. We are made of star-stuff. We are a way for the universe to know itself."

—CARL SAGAN

Star-gazing is mesmerizing. As a child, I was constantly enthralled by the beauty, vastness, and mysteries of the night sky. This led to an intense curiosity to understand the dynamics of the universe, the solar system, Earth, and who we might be within these boundaries. As a species, human animals are acutely inquisitive. We, in general, have an insatiable appetite to know and understand. The motivations to embark on quests for knowledge and understanding are mixed. At their best, these appetites to know are motivated by intense desires to comprehend, realize, and appreciate the breadth, depth, dynamics, and limits life as we encounter and experience it. Although deceptively facile to say, human quests for knowledge and understanding influence principles and strategies about how to live, and for what to live. This blend of curiosity, a need to comprehend, and to create ways to live, is at the heart of some of the most significant and classic human quests: Where we are? Who we are? How are we to live? What is our role in the scheme and parameters of reality?

For some people, like myself, these questions were vital at a very young age. I was not involved with religion, and became deeply molded and psychically shaped by growing up on the shores of Lake Huron, spending years contemplating the waves, being mesmerized by the natural world, and entirely captivated by the night sky. These coalesced as a young adult to the degree that little else was of interest but the quest to know from where, why, and how all this reality came to be. I realized later that these were of the order of archetypal religious, numinous, and liminal experiences. Throughout this time were recurrent experiences of suffering, which also became fastened to my psychic architecture. The power of such existential encounters provoked a deep awakening, a thirst for clarity, and a distain for superficial answers. These themes have consistently oriented and ordered my life.

The consequences are a strong desire for intelligibility and coherence, inclusive of profound sensitivities towards and appreciation of the natural world, a need to comprehend the cosmos, joined with an attentiveness to suffering and a thirst for justice. My journeys required intense studies of theology and religion, religious experiences, science and evolution, liberation and feminist analyses, epistemology, aesthetics, and a plethora of ongoing inquiries that contribute to these quests. I have been privileged to spend years in L'Arche in the community with Jean Vanier, to regularly learn from and converse with Thomas Berry, to study with Gustavo Gutiérrez in Peru, and to be engaged with vibrant and challenging feminist, ecofeminist, ecological, and religious communities of scholars and activists. Consistently, I am drawn to thinkers who blend science, religiosity, and poesies, and who do not fit customary classifications. It always seemed evident to me that we live in a *divine milieu*, to use the elegant phrase of Pierre Teilhard de Chardin. I cannot prove this, but I would stake my life on it. Hence the quest to live in the cosmos has been, and is, my quest.

Most disciplines and modes of inquiry are involved and implicated in the quest to live in a cosmos. From the earliest writings of the Bronze Age, through the Vedic Sanskrit texts of the seventh century BCE, to 300 BCE and the *Four Books* and *Five Classics* of Confucius and his disciples, to the contemporary investigations into new materialisms, entanglement and emergent complexity theories, the quests endure. Whatever the human animal is, it includes such pursuits. These are my quests, and continue to propel my life and work, with an existential insistence I am hard pressed to explain, or contain.

The quest to understand the cosmos represents many journeys: an outward journey to the boundaries of the universe, Earth's journey, the human journey, and the interior journey of integrating these together. The following reflection pauses on these facets of the human quest to live in a cosmos, using insights from a few mentors, while meandering through the development of symbolic consciousness and the power this provides for such a quest. A final section contemplates the question of how are we to live in a cosmos.

Exterior Quest: Where Are We?

The expansiveness and entanglement of time, space, and materiality are mesmerizing actualities. There is a steady, even unrelenting, lure to apprehend the facets of the "orders of reality." The quest to observe and fathom the exterior limits of reality is evident throughout cultures and human history. These topics—which push the human gaze as far as possible to enable us to see or perceive *where* we are—are the subjects of religious, scientific, and philosophical inquiry over millennia. For some, to grasp the extent of reality has a direct bearing on understanding human existence.

This exterior quest, to apprehend the most comprehensive context and extent of reality, leads to questions about origins, purpose, and metaphysics. Humans look outward, to touch and gauge the boundaries, and perceive the ethos and telos, of reality. The effort to push our consciousness outward, to seek understanding of the vastness and intricacies of time, space, materiality, dark matter and gravity, of quarks, black holes, quantum dynamics and planetary processes is an intense quest for knowledge and meaning.

The resulting answers about the ultimate dimensions of reality transform into the horizon from which we interpret human existence. Historically and currently, multiple and contradictory interpretations abound. The gains in seeing into aspects of the universe, in considerable detail, challenges many cultural or religious origins stories, basic presuppositions of time and space, as well as conventional scientific methods and mechanisms of interpretations. Yet, regardless of the current debates and the myriad worldviews, the intensity of this exterior quest persists.

Christianity, while infinitely varied and divergent, has participated in this "exterior" quest. Thomas Aquinas, Hildegard of Bingen, Pierre Teilhard de Chardin, Alfred North Whitehead, and many others were preoccupied with this pursuit. They considered that one of the exigencies of life, even a spiritual impulse, was to understand the larger parameters of reality. Many of these thinkers pondered questions and offered answers to queries that others have not yet asked.

Thomas Berry can be understood, in part, to be within this lineage.[1] For Berry, the expansiveness and essence of the world—cosmos, Earth, time,

1. Berry, *Dream of The Earth*. For a full bibliography of Thomas Berry's works see http://thomasberry.org/life-and-thought/bibliography. See also Eaton, ed., *The*

space, and processes—were central to knowing anything meaningful about being human. Over his lifetime Berry's predominant concerns became the future of the Earth community. For Berry, to understand anything, one has to know the sequences in the phenomenal order that led to the concern. In order to respond to the ecological crisis, which is planetary, it is necessary to know the history, origins, and dynamics of planet Earth. Earth has its origins in the dynamics and processes of the universe. At first blush it is not obvious how responding to the ecological crisis is best understood by first considering cosmology. However, I hope this will become clear along the path of the quest to live in a cosmos.

The universe—the farthest realm of the exterior quest—is understood with increasing clarity. Scientific modes of inquiry progressively detect the dynamic processes, interconnections, and expansions of the universe. What is increasingly astonishing is that everything about the universe is so much more than assumed or imagined previously. Evidence abounds about the complexities, diversifications, and the development sequences of transformations, and the intricacies and interrelatedness of the emergent universe. It is also increasingly apparent that in spite of a capacity to parse this quest into different physical processes—subatomic physics, astrophysics, nucleosynthesis, dark matter and dark energy, planetary formations, etc.—there is coherence. The universe is integral: unified without being uniform. There is a cohesiveness within the astonishing diversity found in how the universe functions, including in the birth and death of stars, and galaxy and planetary formations.

Overall, in these vast exterior realms of reality, there are patterns, processes, developmental sequences, transformations, evolutions, intensifications, and complexifications. For example, in the transformation from the atomic to the molecular structures, a further degree of intensity develops in these new physical arrangements. One could say reality complexifies. Furthermore, it is entangled. Although each discipline explains particular processes in discrete separated categories, if we step back, it is evident that they are interconnected and interdependent processes. How could it be otherwise? Thus scientists are using terms such as emergent complexity, entanglement, coherence, correspondence, congruence, or intelligibility to describe the overall coordination within the universe.

Coherence and integration are also seen in the evolution and functioning of the biosphere. The biosphere is best described, and explained, with interrelated processes, networks of connections, correspondence, mutual influences, and communication from the molecular and cellular

---

Intellectual Journey of Thomas Berry.

to the planetary processes. What is now known of the intelligibility across time, space, materiality, Earth, and the biosphere is a novel revelation about "where we are."

The success of these quests within the last century has revealed new knowledge, and staggered human notions about the extent of, and dynamics within, the universe. This includes observing massive macro-properties, and discerning infinitesimally micro properties, from supernovas to quarks. A 2013 study, using light wavelengths, indicated that there are 225 billion galaxies in the observable universe.[2] In 2016, South Africa's radio telescope, MeerKAT, detected 1,300 galaxies in tiny corner of universe where only seventy were previously known, at a quarter of its eventual capacity.[3] More intriguing is that approximately 68 percent of the universe is dark energy. Dark matter makes up a further 27 percent. "The rest—everything on Earth, everything ever observed with all of our instruments, all normal matter—adds up to less than 5% of the universe."[4]

These observations make up one dimension of exploration of the cosmic parameters of which Earth is one tiny planet, in a small solar system, in a medium-sized galaxy, in a universe with inestimable billions of galaxies, dominated by dark matter and dark energy, within an expanding fabric of space and time of approximately thirteen billion years and counting. These facts are only answering the question, what is out there? Other modes of inquiry are probing the dynamics of matter, energy, black holes, atomic structures, expansions and collapses, space-time, and origins and destinies of matter and energy within the observable universe.[5] Some of the most scientifically conversant, and challenging to interpret, are from quantum mechanics and the affiliated fertile field of new materialisms.[6]

This "exterior" quest, in its cosmic form, has increasingly revealed that we, as *homo sapiens*, have almost no clue about the dynamics, dimensions, and boundaries of the cosmos. The more the cosmos is studied—with increasingly sophisticated tools, and the ability to discard an interpretative theory if contradicted by new data—the more extraordinary, bizarre, intricate, intriguing, and mysterious it becomes to the human mind. The

2. http://www.skyandtelescope.com/astronomy-resources/how-many-galaxies/.

3. "South African super-telescope reveals distant galaxies and black holes," *The Guardian*, (July 17th, 2016). https://www.theguardian.com/science/2016/jul/17/first-image-from-south-african-super-radio-telescope-far-better-than-expected.

4. "Dark Energy, Dark Matter," NASA, https://science.nasa.gov/astrophysics/focus-areas/what-is-dark-energy.

5. I am not including the more speculative realms of parallel and the astronomical theories of the multiverse, or multiple universes.

6. Barad, *Meeting the Universe Half-Way*.

investment of time, money, energy, and expertise is indicative of the power of the quest.[7] *Homo sapiens* are drawn to know the cosmos: the attraction is compelling, alluring, and even irresistible. The cosmos provokes the human mind to focus with intensity, purpose, and perseverance to *know* the universe, to encounter it.

Once such a statement is made, the conversation could be seen to move into the realm of subjective musings, poesies, creative imaginings, reveries, or nonsense. I leave the reader to decide. However, after considerable reflection over decades on these topics, I fully embrace the epistemological quagmires that various postmodern analyses have demonstrated about the social construction of thought and knowledge, and of power, persons, politics, and place. The investigations into worldviews or social imaginaries further support this imbroglio, with their own specifications. The many insights from new materialisms thicken the epistemic stew by entangling the processes of knowing and knowledge with multiplicities of materiality, quantum mechanics, and interpretive pursuits. The acumen from the fields of aesthetics, literary criticism, (eco)poetics, and somatic and cognitive studies further reveals enmeshments of myth, narrative, and symbol with cognition, imagination, perception, emotion, and psychic and social processes. Within this epistemological fluidity there is support to claim an interactive relationship between the cosmos and the human. I am not insinuating that the cosmos is seeking the human in any intentional or overt manner. I am suggesting that there is a bond, a coherence and an affinity within the depths of the phenomenal order that incites and excites the human mind to undertake these quests. In some manner, the cosmos presses on human consciousness. I will attempt to clarify.

Rather than entering the epistemological entrapment mentioned above, I will draw upon discussions about symbolic consciousness. These have informed my views for years and best illuminate the path I am proposing.[8] It requires a momentary meandering from the main path.

7. Of course not all humans are captivated by these quests, and most are unable to begin to embark on these due to hosts of social ills and structural oppressions that cripple the lives, minds, and hearts of many. While cognizant of the critique that it is only the affluent who can undertake such quests, my emphasis is on the undeniable fact that this quest to know the boundaries of the cosmos, even if embarked upon by a few privileged persons, is a constant in human history.

8. I have published frequently on this topic, and am drawing from some of this work. For example see Eaton, "A Spirituality of the Earth"; "The Challenges of Worldview Transformation"; "Forces of Nature"; "An Ecological Imaginary"; and "Insights from Evolution, Cosmology and Earth Sciences."

Symbolic Consciousness and the Quest to Live in a Cosmos

Within the evolution and development of the hominid species emerged the capacity to navigate the world symbolically and then to live by means of a symbolic consciousness.[9] It is this mode of consciousness that allows for representations of the world to form, eventually as worldviews or social imaginaries. There are a plethora of studies on the stages of this development. The increased subtleties of memory enabled representations of perceptually similar episodes, which led to abstractions, and to analytic and associate modes of thought. These processes led to the formation of worldviews. For many years Diederik Aerts has focused on the emergence of worldviews, which he describes as a self-modifying, integrated internal model of the world.[10] Unfortunately, his work remains largely unknown in North American studies of consciousness and worldviews.

The evolution of a self-reflexive consciousness that could function symbolically and sustain the capacity to coordinate images, thoughts, emotions, intuitions, and insights developed over millennia. Many animals exhibit the similar capacities in their communication or language system. In primates and later in *homo sapiens* these faculties became intensified. Research into the modalities of symbolic consciousness—the affective dimensions, and the complex interrelations among symbol, language, emotion and thought process and worldviews—is fascinating albeit provisional. It is now thought that the formation of symbolic consciousness and the capacities of representation are older, more complex, and involves more species than previously speculated.[11] Still, the exact processes that led to communication, signs, representations, art, and imagery—all foundational to language and symbolic consciousness—remain opaque.

9. This understanding of symbolic consciousness comes from several sources: Deacon, *The Symbolic Species*; Dixon, *Images of Truth*; van Huyssteen, *Alone in the World?*; Lewis-Williams, *The Mind in the Cave;* Greenspan and Shanker, *The First Idea*; and Pfeiffer, *The Creative Explosion*.

10. See *The Worldviews Group* which consists of Diederik Aerts (theoretical physics), Bart De Moor (engineering sciences), Staf Hellemans (sociology), Hubert Van Belle (engineering sciences) and Jan Van der Veken (philosophy). Their web site lists some of their publications: http://www.vub.ac.be/CLEA/dissemination/groups-archive/vzw_worldviews/. They also publish in other groups on aspects of worldviews. For a list of publications on worldview see David Naugle's web page: http://www.leaderu.com/philosophy/worldviewbibliography.html.

11. See Klein and Edgar, *The Dawn of Human Culture*. There is little agreement on when, where, how and which version of hominids began to manifest creative and symbolic thinking. See also Wilford, "When Humans Became Human."

Tool use and tool making indicate distinct processes of consciousness, as corroborated by animal studies and ethography. The use and development of tools requires the capacity to imagine, and indicates a nascent form of symbolic consciousness. Otherwise the rock is just a rock, and the stick, well, a stick. Major evolutionary developments over millennia occurred prior to hominids becoming capable of representing, or documenting experiences (ideation, emotions) in markings or images. As suggests John W. Dixon, even the faint shadow of images and artifacts reveals that experiences were transmuted into a system of images to cope with and demarcate the exigencies of life.[12] Dixon postulates that it is within the move from consciousness to self-consciousness that the intensification of symbolic psychic structures took place. There is a complex weave, not fully understood, among active imaging—imagination—experiencing the world, and a symbolic rendering of the experiences. This symbolic mode of being is the *modus operandi* of humans. Symbolic consciousness is the human process of experiencing and navigating the world. Although the observation seems straightforward, the implications are not. It is not through or with symbols or images that we think and comprehend. It is within symbols. Everything pertaining to human existence is represented, articulated, communicated, coordinated, and possibly even experienced within symbols. Humans are a symbolic species.

The passages from symbolic consciousness to representation and then to worldviews occurred in evolutionary stages, and were codified in experiences that became transferable and transmittable. The dynamics of symbolic functioning can be dissected into aspects, and debated as to how these evolved and when, and were inter-related to which other facet. Yet to be able to dissect worldview elements, formation, and transmission and their embeddedness in aspects of culture and context, reveals little about the dynamics of symbolic consciousness. Even the focus on internal processes that interact with and interpret experiences, emotions, cognition, and ideation, which lead to representation, and all of which infuse identity formation and a sense of self, still renders a superficial, even false, understanding. These facets are interdependent and interinfluential in fluid exchanges. Activities, contexts, events, and symbolic processes are inseparable, interwoven and enmeshed within the very structures of human consciousness and behaviors. Furthermore, they operate within an indivisible personal and social weave. Such human processes are increasingly of interest, and there are many avenues into discerning the functional dimensions of symbolic consciousness and activities.

12 Dixon, *Images of Truth*, 49.

The recognition that worldviews are more than cognitive maps is very important. Many researchers address the concept, content, or architecture of a worldview.[13] They neglect to consider the evolutionary processes and internal dynamics that were acquired and adapted such that humans live within symbols and worldviews, not using symbols. Furthermore, symbolic consciousness is a more nuanced and multi-faceted phenomenon than cognitive processes. Worldviews are an external manifestation of internal personal and social symbolic processes that defy a precise portrayal. Current work on mind-brain associations, imagination and cognition interactions, consciousness, somatic studies, language acquisition, and biosemiotics are addressing this aspect of humans as a symbolic species. I think it is more fitting to name ourselves *homo symbolicus* than *homo sapiens*.

The interaction between symbolic consciousness and the natural world is also a rich terrain of study. Some suggest that the rapport between Earth from which *homo sapiens* emerged and the development of symbolic self-consciousness is the key to the origins of *homo symbolicus*.[14] I think that the interactions with the natural world were, over time, the impetus or driver of symbolic consciousness. Survival depended on symbolic mechanisms and performance as humans expanded their capacity to navigate the exigencies of an unmitigated dependence on the natural world. This included myriad threats to survival, and unusual vulnerabilities as an animal with prolonged childhood dependencies and without defense mechanisms of teeth, claws, horns, antlers, armor, speed, poison, or camouflage.

Human symbolic consciousness and the natural world would have had to be extensively entangled. The consistency and depth of relationships between humans and the natural world, over millennia, would have been intense. Throughout history countless poets and philosophers have referred to the power of the elements. The connection between these and the formation of symbolic consciousness is less explored. Gaston Bachelard is one who pondered and articulated the deeper interior dynamics and correlations.[15] He studied physics, philosophies of science, epistemologies, and psychoanalysis. Bachelard was interested in the intersection of materiality

13. Other terms include social consciousness, cognitive or reality maps, big picture thinking, cosmology, sacred canopy, social or ecological imaginary, and undoubtedly others. For examples see Smart, *Worldviews*; Castoriadis, *The Imaginary Institution of Society*; Clark, "A Social Ecology"; Cummings Neville, "Worldviews."

14. See the works of Berry, *Dream of the Earth*; Dixon, *Images of Truth*; Lewis-Williams, *The Mind in the Cave* and *Inside the Neolithic Mind*; and van Huyssteen, *Alone in the World?*

15. This stellar book describes in depth how humans interact with spaces via the imagination, symbolic consciousness and interiority. Bachelard, *The Poetics of Space*.

and consciousness, or *material imagination*. Of interest here is his claim that the elements of Earth, air, water, and fire became psychodynamic structures that influence the vagaries of consciousness, imagination, reveries, and representations. He suggested that there is a fusion between these elements and the human imagination to such a degree that it is a passionate liaison. Bonds were forged between materiality, imagination, and affectivity that, for Bachelard, precede knowledge. Furthermore, creativity stems from these primal dynamics.

Caves, vistas, storms, seasons, other animals, and the elements of air, water, fire, and Earth formed human sensibilities, consciousness, and self-consciousness. Humans had no *techne* to control, and minimal ability to distance themselves from such powers and immensities. Until very recently the natural world has been the dominant influence on human evolution, history, and development. The intensity of this rapport would eventually evoke representation. In fact, Earth symbolization is considered to be the earliest systematic representation of the world.

Potent experiences of the natural world continue to induce a blend of material, mythic, emotional, and psychic facets that require mediation, representation, and expression. For those who are present to the natural world—the night sky, ocean waves, or to the immensity of "nature"—something can occur which can only be described as an awareness of a kind of presence. It is not enough to "go into nature." One has to be somewhat overtaken by the natural world: its presence, vitality, power, or beauty. Outer and inner landscapes blend, and shape perception of the vibrancy of this natural world. For me, watching the night sky becomes a contemplation, which opens my awareness to something I can only describe as depths of reality that are not immediately apparent. It is something of a liminal experience where the cosmos becomes awake in the depths of my being. I experience this as a living cosmos.

For example, experiences of caves are often described in terms of intimacy, intensity, envelopment, or interiority. The experience of "the immensity of the forest" is common, yet is a multilayered and perplexing interior involvement. Bachelard devoted much of his life to analyzing such occurrences. He perceived that it is an immensity felt while in the forest and described as "of the forest," but experienced within our self-consciousness and our bodies. We experience, and thus imagine and interpret, the forest as radiating other dimensions than the material. We feel this immensity *within* ourselves, although describe it as *out there* in the forest. This immensity of the forest becomes intimate: an energy, potency, or presence pressing in on human consciousness. He called this an intimate immensity. For Bachelard, the imagination and associated symbolic expressions are able to enlarge indefinitely the images and sensations of immensity. Thus the experiences increase in their interior presence and power. The ascribed (often understood as derived) meaning is entangled with our emotional, imaginative, symbolic, and cognitive apparatus, which then propels our responses. If these experiences are interpreted as revealing something mysterious or sacred—the place, presence, and activity—it is perceived to emanate a sense of "otherness," and affirmed as such, within us. We describe these experiences as something mysterious or eternal, of losing oneself and going deeper into a limitless world.

Bachelard proposed a *material imagination* to show the dialectic between the material realm, the unbounded elasticity of the human imagination, and the schemas of interpretations and expressions. Bachelard found that affectivity, rather than the less impressive intellectual or cognitive activity of material experiences, were more compelling or foundational to human knowing. He wrote: "It is not *knowledge* of the real which makes us passionately love it. It is rather *feeling* which is the fundamental value."[16] In a similar vein, Charles Peirce wrote about the rational, progressive, and instinctual dimensions of the mind.[17] Pierce claimed that the rational is the most recent in evolutionary development and hence the most immature. The instinctive impulses, sentiments, dreaming, imagination, memory—the community of passions—are the more mature. The rational mind requires this community of passions for optimal functioning. Parallels can be drawn to the immeasurable interconnectedness affirmed within new materialism, and what Jane Bennett calls a "vibrant matter" that affirms scientifically these insights and intuitions.[18]

From symbolic consciousness to worldview is another step. Given that humans live within symbolic renderings of the world, the many modalities or facets of worldviews tend towards a coherent imaging of self, life, and world. Every culture develops a representation of the world as intelligible and coherent. The form is as a narrative. There is accumulating evidence that humans, individually and collectively, generate and live within narratives: that narrative is the "information and navigation" structure of the mind. This is, of course, not a new idea; however, there seems to be new evidence to support it. Jonathan Gottschall makes a compelling case that humans are "the story telling animal."[19] With verifications from evolutionary biology, psychology, and neuroscience, he shows a multiplicity of ways

16. Quote in Kaplan, "Gaston Bachelard's Philosophy of Imagination," 4.

17. Halton, "Eden Inverted."

18. Bennett, *Vibrant Matter.*

19. Gottschall, *The Storytelling Animals.*

in which humans are always living within and reconstructing experiences in narratives.[20] These narratives are the cognitive, communication, education, and classification mode of humans, as a species. This storytelling mind seeks coherence. It is "allergic to uncertainty, randomness and coincidence. It is addicted to meaning. If the storytelling mind cannot find meaningful patterns in the world, it will try to impose them."[21] Such narratives are not subject to the categories of fact or fiction: they are worldviews. Furthermore, story is the epicenter of individual and social cohesion: "story is the counterforce to social disorder, the tendency of things to fall apart. Story is the center without which the rest cannot hold."[22]

A further point about symbolic consciousness is important. Symbols not only exteriorized and codified experiences. They allowed humans to transmit experiences and accumulate knowledge. This empowered humans to move quickly, from an evolutionary standpoint, from consciousness to a symbolic and self-aware consciousness. The result of which amplified the powers of consciousness, imagination, and creativity. Such systems are referred to as positive feedback or self-amplifying loops: consciousness fashioning symbols that in turn magnify consciousness.

These symbolic mechanisms of the human mind are the bases of knowing. They are the powers through which the mind is extended. It is through symbolic formations that worldviews develop: informed by experiences and formed as stories nested in stories, and infused with meaning, emotion, and values. The interior landscape of navigating and interpreting life engenders symbolically linked activities, which are embedded in social narratives, and shaped into systematic symbolic and collective representations of the world. This empowered and emboldened what was, a few hundred thousand years prior, a vulnerable, weak, and inconspicuous species. It is this same species that has extended these same powers of symbolic consciousness into the farthest reaches of the phenomenal order, back in time 13.7 billion years, probed materiality to apprehend atomic structures, and detected black holes, and invisible dark matter.

What is the impetus for the human quest to know the cosmos? Reasonable answers include that humans are explorers with an insatiable curiosity. Or humans are an animal who refuses to be confined, and is determined to push back boundaries. NASA claims that there is an innate desire to expand

20. Ibid., 99. Gottschall makes a further case that humans of all ages prefer fiction to fact. As well, the storytelling brain, especially the left hemisphere activities, will fabricate a story rather than leave something unexplained, as seen with spilt brain subjects.

21. Ibid., 103.

22. Ibid., 138.

human presence into the solar system and beyond.[23] Others could claim it is a need to prove an omnipotence of sorts, or an extension of domination. While these are sensible options, I am interested in a particular line of inquiry that combines the insights of symbolic consciousness with the quest to encounter the cosmos. The potency of self-consciousness and the elasticity of symbolic consciousness are both the tools and the force of this quest. They also generate the receptiveness for the encounter.

More is known about the boundaries of exterior realities than ever before. One outcome of making this knowledge public and accessible is to amplify and intensify human consciousness about "where we are." Thus humans are becoming more conscious of the universe and Earth. This expansion of awareness strengthens acuity of both the scheme of things, as well as who we are in this scheme of things. Of the many insights I learned from Thomas Berry, one is that this quest to know "where we are" begins with the universe, as it has to begin at the origins of time and space. A second is to develop an incisive attentiveness of how to interpret what we perceive and learn.

The cosmos cannot be seen as a backdrop to the human drama, or as a context, an unfolding, a progression, or a potential. It is not like an embryo that matures into fullness. It is more a becoming: not linear and determined, more so creative and dynamic yet seemingly with an orientation. As the universe develops, it becomes more: more complex, interactive, entwined, vibrant, and intense. That is why, for Teilhard de Chardin and Berry, the best image is that the cosmos is a cosmogenesis. This implies forms of continuity and coherence between cosmogenesis, geogenesis, and biogenesis. In the same manner, evolution is a process or dynamic of the biosphere from which *homo sapiens* evolved, with a form of self-reflexive symbolic consciousness that is able to perceive that these forms of genesis are ongoing. There is coherence and continuity.[24]

What Berry realized is that this is radically new knowledge. The reference points for understanding the universe, Earth, ourselves and our role within the scheme of things all change with this new knowledge. To understand anything, we need to grasp, even at a basic level, that the universe is a primary source and reference. Everything about Earth evolved and developed from cosmic processes. All aspects of *homo sapiens* evolved and developed from Earth processes. To say that Earth formed or produced us is inadequate language. We emerged from and are a conscious living part of

23. NASA, "Beyond Earth."

24. This is not uniformity, or intelligent design where the configuration was predestined.

Earth realities. By extension and extrapolation, the most apt description of the universe is that it is alive.[25]

In this vein, the expansion of human consciousness into the cosmos is also the universe and Earth becoming conscious in humanity. Put differently, it is the universe reflecting on itself in human form, or that humans are a mode of self-consciousness of the universe. To understand and integrate that we are a self-conscious element of a living cosmos is a great challenge. The quest to live in a cosmos is thus a dynamic of the cosmos, and encounter with the cosmos. One way is to enter into this interpretive zone is to see how this exterior quest to know the cosmos is also an interior quest.

Who Are We? Exteriority Becomes Interiority

Of the myriad ways to broach this topic, the basic point is an extension of the above: the quest to understand the largest parameters of reality is intimately involved with "who we are" in the scheme of things, and also with the interior modes of knowing. Of the countless ways to consider interiority, I have two comments related only to this essay's theme. The first remark is about a fusion between exteriority and interiority. Here I am indebted to and influenced by significant aspects of the thought of Pierre Teilhard de Chardin. In terms of interiority, Teilhard contemplated an intimacy between the *without* and the *within of things*.[26] The without is the observable. This includes the structures and changes from the establishment and bonding arrangements of atomic structures, to the formation of molecules and mega molecules out of which arose and evolved all matter. The starting point for understanding the without is the discernible atomic structures and behaviours. Teilhard sought, and developed, a theory that connects structure and activity with processes and purposes of the developmental transformations. He pondered these as a whole, meaning he would not separate anything from its structures, activities, developments, and directionality.

For example, Teilhard studied bacteria cultures in this manner, and then plants. He explained that for plants, the without and observable cannot explain the life dynamics of plants. With insects it is more difficult, with vertebrate futile, and then breaks down completely with humans. As life evolved, the without of things—the observable—becomes increasingly

25. A magnificent exposé of this understanding is in Tucker and Grim, eds., *Living Cosmology*.

26. This concept is introduced and developed throughout Teilhard de Chardin, *Le Phénomène humain* and *The Human Phenomenon*.

incapable of explaining behavior, development, intensifying complexity, and evolutionary directionality. Some form of interiority, "within of things," *élan vital*, vitality, subjectivity, *Geist*, *ch'i* is increasingly present, active, effective and indeed essential, everywhere. This within of things, its interiority, is a subtle, nuanced union of matter, energy, spirit, and telos that coheres the interior dynamics with the transformations to increasing levels of complexity. The within of things is manifested in the overall orientation and processes that compel atoms to transform to molecules . . . to form planets, Earth, an atmosphere and biosphere, to life, consciousness, and self-consciousness. Herein we see something similar to the congruence, intelligibility, and coherence mentioned above.

Teilhard used the term *consciousness* with many qualifications. Consciousness itself is differentiated and evolves, from nascent traces to that of life and mind. Inert matter does not have consciousness *per se*, although through his lengthy discussions of the within of things, Teilhard maintained that something interior, not observable yet clearly present, moves the process of evolution. Overall, in *Le Phénomène Humain*, he wanted to write the natural history from the without and the within of things, which combined spirituality and science. This was predictably contentious for both disciplines. For me, however, it was cogent, brilliant and beautiful, and reverberated deeply with my quest.

There is a further aspect to Teilhard's thought that is important to appreciate in his elegant blend of science, religiosity, and poesis. Each stage of evolution is nascent in the previous, but not in a simple embryonic or potential form. There are critical changes that alter the very ontology of reality. For example, in the transformation from the atomic to the molecular structures, the new arrangements of the parts required the acquisition of another dimension—a further degree of interiority—that allowed reality to complexify. Every evolutionary development requires an intensified and differentiated structure that corresponds to a more subtle and supple concentrated interiority and consciousness. Scientifically, reality thickens, deepens, and crosses new ontological thresholds. Spiritually, interiority intensifies and amplifies.

Berry concurred that there must be differentiated forms of interiority within the processes of the universe, Earth, and the biosphere. These dynamics are the creative energies interior to all sequences of transformations. These relationships between the without and within of things, between exterior and interior dynamics and processes, occur at all times and are intensified at every development phase. Berry used the language of differentiation, subjectivity, and interrelatedness to convey a similar insight. What is relevant here is that many of the intuitions that Teilhard proposed are being verified through science, although described with different language and not interpreted in a spiritual framework.[27]

A second remark is about interior awareness. How are we to absorb these new findings about where and who we are? Even the most hard-nosed evolutionary scientists must accept that life emerged from Earth dynamics: from the interior of cosmic and planetary processes. By extrapolation, the same must be said of consciousness, self-consciousness, and symbolic consciousness. Therefore, it is logical to claim there is an emergent interiority: highly differentiated among species. Yet while the observations and logic are difficult to deny, the implications are far reaching and do not fit neatly into most operative worldviews.

To study evolution is to realize that the biosphere thrives in integrated and interdependent relations, from the interwoven atmospheric, climate, and water systems to fractal patterns and cellular dynamics. The complexity and ingenuity of Earth processes such as self-organizing dynamics, natural selection, emergence, symbiosis, and co-evolution become apparent. Earth enlivens interconnected webs of bacteria, insects, plants, animals and their related social patterns, and forms of consciousness. To attend to evolution, even minimally, is to be dazzled. Earth's intricacies animate the human imagination. The immense and elaborate planetary hydrologic cycle is stunning and breathtaking. From the microbiotic and genetic levels to the dinosaurs, the processes and life-forms are astonishing. To see the elegance of birds, the ingenuity of insect communication, and the emotions of mammals is to be thrilled and overwhelmed by creativity, diversity, power, and beauty. These intimate immensities nourish human depths, or at least they could.

All animals, including humans, need first to be understood as differentiated yet integrated living elements of a whole. To grapple with the implications of evolutionary complexities propels a momentous perspectival shift. Elsewhere I have described this as the revolution of evolution.[28] It is my view, and experience, that becoming aware of the extraordinary dynamics of evolution can open up the possibility of profound depth or religious experiences. Such experiences allow a glimpse into a world of stunning elegance, of mysteries and adventure, of vistas beyond our knowing. The natural world inspires wonder and awe: a kind of power available to all who attend carefully to the natural world. The movement of the stars, the presence of mountains, the invigorating quality of ocean waves fills us with

27. For example see Margulis and Sagan, *Dazzle Gradually*; Margulis, *The Symbiotic Planet*; Goodenough, *The Sacred Depths of Nature*; and Sahtouris, *Earthdance*.

28. Eaton, "The Revolution of Evolution."

feelings of celebration and reverence. The eloquence of Abraham Joshua Heschel is worthy of a pause:

> Awe is an intuition for the dignity of all things, a realization that things not only are what they are but also stand, however remotely, for something supreme. Awe is a sense for the transcendence, for the reference everywhere to mystery beyond all things. It enables us to receive in the world intimations of the divine . . . to sense the ultimate in the common and the simple; to feel in the rush of the passing the stillness of the eternal. What we cannot comprehend by analysis, we become aware of in awe.[29]

We are moved, like Teilhard de Chardin, to claim we live in a divine milieu and that matter, spirit, and life are intertwined in a sacred process. We can see a deeper reality: one that kindles the imagination, awakens us to the Earth, and ignites a fire and desire to protect the biosphere. Familiarity with evolution can open awareness or consciousness to Earth mysticism, a blend of the best of science and religion.[30]

This kind of description is uncommon, and is unacceptable in many academic discourses. It is too subjective, too emotive and imprecise. It becomes a kind of ecopoeisis, and then is readily dismissed. We are accustomed to scientific, philosophical, ethical, and aesthetic modes of inquiry to be separate, and for academic parlance to be verifiable and solemn. Attributed to the great Mark Twain is this apropos comment: *The researches of many commentators have already thrown much darkness on this subject, and it is probable that if they continue we shall soon know nothing at all about it.*

Our habitual modes of thought and language categories, especially in academia, are inadequate for this synthesis of knowledge, insights, and affectivity. Hyper-rational modes of inquiry are restrictive. Customary intellectual tools that measure, define, analyze, critique, and deconstruct hegemonies have limits. These intellectual processes, while valuable, neither come from nor speak to the depth of human interiority. Hence they cannot illuminate what is being learned of the comprehensiveness and coherence of the universe. My preferred mentors (Bachelard, Teilhard, Berry, Heschel) had a great appreciation for many modes of knowing and perceiving, including dreams, stories, imagination and *poesis.* Other sensibilities, such as emotions, intuitions, insights, presentiments, wonder, and wisdom are valid

29. Dresner, ed., *I Asked for Wonder*, 3.

30. Two unrelated Kaufmans have written on these topics. Stuart Kauffman, *Reinventing the Sacred* and Gordon Kaufman, *In the Beginning.*

indicators of knowledge. Interiority and depth perceptions are a place of great vitality, elasticity, and inventiveness.

The human pursuit to understand the cosmos is both an exterior and interior quest. There is an interior aspect to the dynamics and processes, and it is in interiority that we experience the universe. Those who quest to live in a cosmos, experience the cosmos. The intimate immensities of the cosmos are perceived or intuited within these interior expansions of symbolic consciousness of the self. It becomes evident, over time, that this is an encounter. It is possible to learn the data and remain unmoved, but as Bachelard would point out, there has thus been no passionate liaison, no affectivity. The material imagination is not engaged, and only the inferior functions of the intellect, cognition, and rationality are involved.

There are myriad continuities between the breadth of knowledge of the living universe and a depth of inner awareness. This new knowledge expands interiority, and magnifies consciousness. If we can absorb it, our horizons enlarge, our awareness heightens, and our religious sensibilities intensify. Again Heschel says it best:

> We can never sneer at the stars, mock the dawn or scoff at the totality of being. Sublime grandeur evokes unhesitating, unflinching awe. Away from the immense, cloistered in our own concepts, we may scorn and revile everything. But standing between earth and sky, we are silenced by the sight.[31]

How are we to live? To live in a cosmos

It was in these exterior and interior quests that Berry saw a way forward to respond to the ecological crisis, with a transformed cultural orientation. His intellectual acumen as a historian of religions and culture, and his astute awareness of religious experiences, poesies, scientific knowledge, and so much more became embedded in his dream of the earth and cosmological proposal. This was not for the purpose of expanding knowledge and consciousness. He shaped and interpreted the knowledge to activate dimensions of interiority to respond to the escalating ecological and social crises.

Our cultural and religious maps are not functioning in the interests of a vital biosphere. How are we sure? Look at what is happening to Earth. Look at our economic systems, at the escalating violence, the war on terror, the war on women, consumerism, and billion dollar arms industries while

31. Dresner, ed., *I Asked for Wonder*, 2.

people starve or have no health care. Anthropogenic climate change is now recognized, but action plans are gridlocked among competing interests and powerful lobbies. Post truth, fake news, and alternative facts are dulling human sensibilities and shrinking inner and outer horizons of meaning. There is much discussion about why we cannot move effectively on ecological issues.

We cannot perceive an adequate orientation towards the planetary demands of the present. For Berry, responses lie within the cultural visions, social imaginaries, or stories. The current versions are dysfunctional in their larger social and ecological dimension, and are not providing direction for a viable future. What stories could give guidance for our era? What gives us an exterior and interior orientation to integrate the most we can know about where we are, and who we are? The response here is that it is the universe and Earth, in all their complexities, majesty, diversities, and exigencies that educates and orients the depth sensibilities of the human animal. Religion and science need to collaborate to perceive the psychic-spiritual dimensions intimately interwoven in the physical-material. In order to respond to current challenges, we need to appreciate the magnitude and magnificence of existence. To live in a cosmos is to experience it as intimate immensities, which illuminate a path, and radiate radical openness. To live in a cosmos, the emergent universe, as the primary reality, can offer such an orientation. The cosmos is not just "out there." It is also within. If we can discover our role in these larger evolutionary processes, there may be hope.

> *The more clearly we can focus our attention on the wonders and realities of the universe about us, the less taste we shall have for destruction.* —Rachel Carson

Bibliography

Bachelard, Gaston. *The Poetics of Space*. Translated by Maria Jolas. Boston: Beacon, 1964.

Barad, Karen. *Meeting the Universe Half-Way: Quantum Physics and the Entanglement of Matter and Meaning*. Durham, NC: Duke University Press, 2007.

Bennett, Jane. *Vibrant Matter: A Political Ecology of Things*. Durham, NC: Duke University Press, 2010.

Berry, Thomas. *Dream of The Earth*. San Francisco: Sierra Club, 1988.

Castoriadis, Cornelius. *The Imaginary Institution of Society*. Translated by Kathleen Blamey. Cambridge, MA: MIT Press, 1998.

Clark, John. "A Social Ecology: An Ecological Imaginary." RA Forum, April 25, 2015. http://raforum.info/spip.php?article1050&lang=fr.

Connolly, William. "The 'New Materialism' and the Fragility of Things." *Millennium Journal of International Studies* 41.3 (2013) 399–412.

Cummings Neville, Robert. "Worldviews." *American Journal of Theology and Philosophy* 30.3 (2009) 233–43.

Dalton, Anne Marie. "Communion of Subjects: Changing the Context of Questions About Transgenic Animals." *Worldviews: Global Religions, Culture and Ecology* 13 (2009) 1–11.

Deacon, Terrence. *The Symbolic Species: The Co-evolution of Language and the Brain.* New York: W. W. Norton, 1998.

Dixon, John. *Images of Truth: Religion and the Art of Seeing.* Atlanta: Scholars, 1996.

Dresner, Samuel, ed. *I Asked for Wonder: A Spiritual Anthology Abraham Joshua Heschel.* New York: Crossroad, 1997.

Eaton, Heather. "The Challenges of Worldview Transformation: To Rethink and Refeel our Origins and Destiny." In *Religion and Ecological Crisis: The "Lynn White Thesis" at Fifty*, edited by Todd Levasseur and Anna Peterson, 121–36. New York: Routledge, 2016.

________. "An Ecological Imaginary: Evolution and Religion in an Ecological Era." In *Ecological Awareness: Exploring Religion, Ethics and Aesthetics*, edited by Sigurd Bergmann and Heather Eaton, 7–23. Studies in Religion and the Environment, Vol. 3 / Studien zur Religion und Umwelt, Bd. 3. Berlin: LIT, 2011.

________. "Forces of Nature: Aesthetics and Ethics." In *Aesth/ethics in Environmental Change*, edited by Sigurd Bergmann, Irmgard Blindow, and Konrad Ott, 109–26. Studies in Religion and the Environment, Vol. 7 / Studien zur Religion und Umwelt, Bd. 7. Zurich: LIT, 2013.

________. "Global Visions and Common Ground: Biodemocracy, Postmodern Pressures and The Earth Charter." *Zygon: Journal of Religion and Science* 49.4 (December 2014) 917–37.

________. "Insights from Evolution, Cosmology and Earth Sciences." In *ECOTHEE: Ecological Theology and Environmental Ethics*, edited by L. Andrianos, K. Kenanidis, and A. Papaderos, 407–426. Institute of Theology and Ecology, Orthodox Academy of Crete: 2009.

________. "A Spirituality of the Earth." In *The Nature of Things: Rediscovering the Spiritual in God's Creation*, edited by Norman Habel and Graham Buxton, 229–40. Eugene, OR: Pickwick, 2016.

________. "The Revolution of Evolution." *Worldviews: Environment, Culture, Religion* 11.1 (Spring, 2007) 6–31.

Eaton, Heather, ed. *The Intellectual Journey of Thomas Berry: Imagining the Earth Community*. Lanham, MD: Lexington, 2014.

Eaton, Heather, and Anne Christine Hornborg. "Ritual Time and Space: A Liminal Age and Religious Consciousness." In *Religion, Ecology and Gender: East-West Perspectives*, edited by Sigurd Bergmann and Yong-Bock Kim, 79–90. Studies in Religion and the Environment / Studien zur Religion und Umwelt. Zurich: LIT, 2009.

Gottschall, Johathan. *The Storytelling Animals: How Stories make Us Human*. New York: Houghton Mifflin Harcourt, 2012

Greenspan, Stanley, and Stuart Shanker. *The First Idea: How Symbols, Language, and Intelligence Evolved from Our Primate Ancestors to Modern Humans*. Cambridge MA: Da Capo, 2004.

Goodenough, Ursula. *The Sacred Depths of Nature.* Oxford: Oxford University Press, 2000.

Halton, Eugene. "Eden Inverted: On the Wild Self and the Contraction of Consciousness." *The Trumpeter* 23.3 (2007) 45–77.

Kaplan, Edward. "Gaston Bachelard's Philosophy of Imagination: An Introduction." *Philosophy and Phenomenological Research* 33.1 (1972) 1–24.

Kauffman, Stuart. *Reinventing the Sacred: A New View of Science, Reason and Religion.* New York: Perseus, 2008.

Kaufman, Gordon. *In the Beginning: Creativity.* Minneapolis: Fortress, 2004.

Klein, Richard, and Blake Edgar. *The Dawn of Human Culture: A Bold New Theory of What Sparked the "Big Bang" of Human Consciousness.* New York: John Wiley and Sons, 2002.

Kuhn, Thomas. *The Structure of Scientific Revolutions.* Chicago: University of Chicago, 1970.

Lewis-Williams, David. *The Mind in the Cave: Consciousness and the Origins of Art.* New York: Thames and Hudson, 2002.

_________. *Inside the Neolithic Mind: Consciousness, Cosmos, and the Realm of the Gods.* London: Thames & Hudson, 2005.

Margulis, Lynn, and Dorion Sagan. *Dazzle Gradually: Reflections on the Nature of Nature.* White River Junction, VT: Chelsea Green, 2007.

Margulis, Lynn. *The Symbiotic Planet: A New Look at Evolution.* New York: Basic, 1998.

Mills, C. Wright. *The Sociological Imagination.* New York: Oxford University Press, 1959.

NASA. "Beyond Earth." Accessed April 13, 2018. https://www.nasa.gov/exploration/whyweexplore/why_we_explore_main.html#.WM2x4XfMw1g.

Naugle, David. *Worldview: The History of a Concept.* Grand Rapids: Eerdmans, 2002.

Norwine, Jim, and Johathan Smith. *Worldview Flux: Perplexed Values for Postmodern Peoples.* Lanham, MD: Lexington, 2000.

Oelschlaeger, Max. *The Idea of Wilderness: From Prehistory to the Age of Ecology.* New Haven, CT: Yale University Press, 1991.

Pfeiffer, John. *The Creative Explosion: An Inquiry into the Origins of Art and Religion.* New York: Harper & Row, 1982.

Smart, Ninian. *Worldviews: Cross-cultural Explorations of Human Beliefs.* New York: Scribner's, 1983.

Soper, K. "Unnatural times? The social imaginary and the future of nature." *Sociological Review* 57.s2 (2009) 222–35.

Sahtouris, Elisabeth. *Earthdance: Living Systems in Evolution.* San José, CA: iUniverse, 2000.

Sheets-Johnstone, M. *The Roots of Morality.* University Park, PA: Pennsylvania State University Press, 2008.

Swan, Liz Stillwaggon. *Origins of Mind.* Dordrecht: Springer, 2013.

Taylor, Charles. *Modern Social Imaginaries.* Durham, NC: Duke University Press, 2004.

Teilhard de Chardin, Pierre. *Le Phénomène humain.* Seuil: Paris, 1956. Republished as *The Human Phenomenon,* translated by Sarah Appleton-Weber. East Sussex, UK: Sussex Academic, 1999. First published in English as *The Phenomenon of Man,* translated by Bernard Wall. New York: Harper, 1959.

Tucker, Mary Evelyn, and John Grim, eds. *Living Cosmology: Christian Responses to the Journey of the Universe.* Maryknoll, NY: Orbis, 2016.

van Huyssteen, Wentzel. *Alone in the World? Science and Theology on Human Uniqueness*. Grand Rapids, MN: Eerdmans, 2004.

Wilford, John Noble. "When Humans Became Human." *New York Times*, February 26, 2002. Accessed February 8th, 2017. http://www.nytimes.com/2002/02/26/science/when-humans-became-human.html?pagewanted=all.

Author Index

Subject Index